# Managing Maintenance Shutdowns And Outages

Joel Levitt

Industrial Press
New York

Library of Congress Cataloging-in-Publication Data
Levitt,Joel, 1952-
Managing maintenance shutdowns and outages / Joel Levitt.
p. cm
Includes bibliographical references and index.
ISBN 0-8311-3173-X
1. Plant maintenance--Management. 2. Plant maintenance--Planning.
3.Plant Shutdowns.
I. Title
TS192.L474 2004
658.2'02--dc22
2003070041

First Edition 2004
Managing Maintenance Shutdowns and Outages

Industrial Press Inc.
200 Madison Avenue
New York, New York 10016



Text and Cover Design: Janet Romano
Manageing Editor: John Carleo

2 3 4 5 6 7 8 9 10

*Disclaimer: The advice given in this book is general. The judgment of qualified people is essential to determine the fitness of any of these statements for your use. All statements must be adapted to the site conditions of any individual shutdown. The author and publisher take no responsibility for any adverse outcomes.*

# Dedication

I have been blessed with the opportunity to teach maintenance management topics to thousands of maintenance professionals worldwide for universities, private companies, publishers, and software companies. I want to dedicate this work to those folks who sponsor professional development training.

The idea for this work originally came about from a request to give a course in Shutdowns and Outages in Trinidad from Anne DeSousa of Trainmar. Thank you Anne and your able assistants Camille Whitby, Janna Turpin and Neisha Mohammed. The course subsequently was moved to South Asia, sponsored by Partners in Singapore. Thank you Delphine Ang and Rona Lim.

The able folks at New Standard Institute helped in that Mike Brown taught me his approach to shutdowns and planning and Tessa Marquis got me the opportunities to teach. Thank you both.

I have been giving courses for the University of Alabama since 1989 and have enjoyed the connection with this great institution. Thanks to Tom Wingenter and his able staff for running an excellent program. My first sponsor was Clemson University and USC (now called Saddle Island Corporation). The person who observed me in 1987 and thought I'd make a decent instructor was Jean Hey. Thanks wherever you are. I gave over 100 courses for Betsy Tyson at Saddle Island, Thanks.

There were many others in the past that worked diligently to produce good and educationally sound programs on maintenance management. I appreciated the opportunity to represent you and provide the best training that I could in these topics.

Sincerely,
Joel Levitt

# ACKNOWLEDGEMENT

This is the author's sixth work on maintenance management topics. These books would have been impossible without the team behind the scenes that make them happen. I couldn't ask for a better senior editor then John Carleo. From the first book, Managing Factory Maintenance in 1996 until now, John has been completely supportive of my projects and of me.

The look and feel of the book is in the competent hands of Janet Romano, Art Director. She has been great to work with and has turned out a handsome set of books (I'm not at all biased).

I'll bet you thought I knew some grammar and usage. Actually each book has been copy edited by either Robert Weinstein or Bob Green. Thanks to all of you.

I also would like to thank those maintenance professionals whose names appear throughout the book, who lent me (in some cases hard-earned) stories. Learning the lessons from some of their stories could save you thousands on your next shutdown.

# Use of Terms

In this book the terms outage, shutdown, and turn-around are used interchangeably. Different companies and industries favor different terms and the maintenance world recognizes that these words have slightly different meanings. For our purposes they are the same.

This book is designed to cover scheduled events. In this model, all shutdowns, turnarounds, and outages have been approved and discussed and the implications thought through. An issue that comes up is the applicability of these ideas and processes to unscheduled shutdowns (breakdowns are sometimes called forced outages).

The processes described here are designed specifically for scheduled events but they can be applied in abbreviated forms to breakdown or other non-scheduled shutdowns. By all means, apply whatever techniques are useful to gain control of the event, and good luck.

***PM = Project Management:*** In this book PM usually stands for Project Management and not Preventive Maintenance

***PMS = Project Management Software***

**CMMS** = Computerized Maintenance Management Software

"There are only two types of work on a turnaround, routine, and unexpected. If the routine is under control there is time to deal with the unexpected, but if the routine becomes unexpected the unexpected may become catastrophic."

*Tom Lenahan in the Preface to Turnaround Management.*

# INTRODUCTION

**This book is designed for:**

This book is designed for use in heavy maintenance environments including refineries, power plants, chemical plants, mines, airports, large factories, and large facilities. It was designed specifically for the project teams that might consist of maintenance managers, project engineers, supervisors, maintenance engineers, planners, clerks, and people who are in training for these positions. Prior background in shutdowns or Project Management would be useful but is not essential.

Another target audience consists of the many contractors who participate in shutdowns. These groups might be in total turnkey control of the shutdown or might only supply labor for the event.

Of the material in this book, 80% is concerned with the time before the first spade of dirt is moved, 15% is execution tactics to keep the project on course, and the remaining 5% is concerned with reporting and recording the lessons learned.

**Promise of this book:**

Simply, the promise of this book is that you will save money on your next outage. To achieve that saving, we promise you will have a new and deeper understanding of how to effectively manage the myriad of details of large maintenance jobs and projects such as power plant outages, refinery refits, etc. This understanding will include an increased ability to plan for and manage such projects, leading to savings by avoiding problems.

**Three Major Costs**

There are three major cost items in a factory, process plant, oil drilling platform, or other large facility.

1. Capital investment, including building and extending or expanding a plant or facility.
2. Catastrophic breakdowns, forced shutdowns and large accidents, fires, and explosions.
3. Intentional scheduled shutdowns, outages, & turn-arounds.

This course is designed to help an organization manage the costs of the third item, intentional shutdowns, outages, and turn-arounds. The concepts and ideas can also be used to manage both the first item (but the driver and organization is somewhat different), and the non-scheduled event.

This book also includes a complete shutdown glossary and a outside resources section.

**The good old days:**

I was speaking to an old-timer craftsperson from a coal-fired, base-load power plant in the western US about the good old days. He said that, when he was young, if he saw a problem that required an outage he ordered the parts needed to do the work. When the parts came in he put them aside, adding a tag with his name. He told the superintendent what he was going to do a week or so before the shutdown. This method of scope management was repeated by many of the 75 other craftspeople in the plant. Of course this was for smaller jobs; the big projects, even in those days were planned ahead.

# TABLE OF CONTENTS

# INPUTS TO THE SHUTDOWN

CHAPTER 1

# Basics of Shutdowns, Turnarounds, and Outages

## Understanding the need to manage shutdowns, turnarounds, and outages

Project work in general and shutdowns in particular are in conflict with just about everything else going on at any time. If the advantages were not so great, shutdowns would never occur. Projects involving shutdowns of major equipment are expensive in direct costs (money for labor and parts) and expensive in terms of lost production. In some instances, the value of the lost production is an order of magnitude greater than the direct costs of the shutdown itself.

Because of the high combined costs and potential impact (both positive and negative) on the business, shutdowns can have intense top management attention. As a result of the cost, impact, and attention, these events are more intensively managed than other large maintenance jobs. This aspect has both good and bad (from a shutdown management perspective) results. On the good side, resources are usually easier to come by. On the negative side we can see excessive top management meddling. Like military actions, professionals should run shutdowns.

To make the best decisions in preparation for and during shutdowns it is important to understand what makes your business or organization tick. Shutdown projects and project strategies flow from a thorough understanding of what most influences your business (that is, what are the dominant patterns such as growth, cost avoidance, safety, efficiency, etc.). These influences are usually a mixture of external factors that must be forecast or predicted, and internal factors over which you have various degrees of control. It is not rocket science to realize that a scheduled shutdown of an electric utility in the summer would be contrary to the business need.

A very good justification would have to be established before a summer shutdown in that industry to make business sense.

Smaller maintenance projects generally do not need extensive justification. Most projects must be justified (but less rigorously) and the justification for most maintenance and repair work is self-evident.

Major projects, shutdowns, and construction, follow similar processes to those used in planning for shorter maintenance jobs and projects. The difference is in the range of the scope of work, the number of people, the use of contractors, the number of contractors, the amount of coordination, and the level of justification.

Planning and Scheduling for major shutdowns is a specialty. It is generally not practical for the same individual to concurrently plan and schedule normal weekly / daily effort as well as major shutdown efforts. Except in the smallest facility it would be impossible to keep up with both sets of demands. Something would slip.

For example in a small heavy-oil refinery there are two planners: one for all maintenance work and another, separate planner who plans and then manages the shutdowns, major work, and construction.

## Size (of the shutdown) Matters

There are different sizes of shutdowns. Size (in addition to complexity and familiarity) makes a major difference in the size of shutdown management staff, how detailed the planning needs to be, how much oversight is needed, what is the best organizational structure, and every other item in the management of the shutdown. In short size matters!

When we review the statistics from a number of shutdown events from different industries we find that the size can be logically broken into four categories:

The structure of the shutdown and the management requirements will change as events become larger. The largest events become more like building a new plant and are managed like any large complex construction job. The smaller events are handled more like large maintenance jobs without much extra fanfare.

The preplanning time line is greatly impacted by the size and scope of the shutdown. The approximate amount of planning needed will be discussed later, but the formal planning process for a multi-million dollar event might start a few years in advance of the event.

| Size of Shutdown | Dollars spent in Parts and Labor | Duration of shut down | Leadtime | Use of Project Software | Contractors | How is the team constituted |
|---|---|---|---|---|---|---|
| Small | Under $250,000 | Hours | Weeks | Sometimes | Usually lowpercentage | Existing staff |
| Small Moderate | $250,000 to $1,500,000 | Short days | Months | Usually | Usually moderate percentage | Usually existing staff |
| Moderate large | $1,500,000 to $10,000,000 | Days | Years months | Always | High percentage | Usually separate shutdown staff |
| Large | $10,000,000 up | Weeks | Years | Always and intensively | Very High percentage | separate staff for shutdowns needed |

## Why Have Shutdowns?

(Keep in mind that maintenance work done under shutdown conditions is expensive).

One factor in managing shutdowns is the answer to the question why have a shutdown? There are six categories of reasons and they all boil down to improved performance of the plant (in one form or another) or compliance with the law.

One reason for a shutdown is changes in market demand (new products or need for increased capacity).

a. Meet a competitive challenge
b. Meet an expanding market
c. Open a new market

New products are constantly being invented, and old plants are adapted to current needs. Shutdowns are designed to change the plant's output.

Profit enhancement (efficiency improvements to save money or reduce the cost of making what you already sell)

a. Operational efficiency
b. Energy efficiency
c. Reduced scrap or increased yield (major driver of the semi-conductor industry)

Natural gas is expensive but clean, and oil costs go up and down. We refit to improve the return on investment from our physical assets. For instance, modern gold mines in South Africa found a new process that extracts more gold from the ore (in fact the new processes can take the tailings from old gold plants and economically reprocess them to yield more gold).

Maintenance need (replacing worn out assets) is one of the most common drivers for the shutdowns described in this book. Maintenance need might be based on PM inspection (wear plate is shot), NDT (tubes are too thin), history (we rebuild every 4 years), even gut feelings (rarely, but sometimes it is the only way). We need to:

a. Increase reliability
b. Increase repeatability
c. Increase or augment life span

Westinghouse or GE recommends that their power-generating turbines be taken down and serviced every few years. Jaws in crushers wear out and have to be replaced periodically. The scheduled outage is common in the power industry.

Customer's request (process improvement, automation to reduce the number of operators) for increased production might drive a shutdown to:

a. Increase throughput
b. Increase quality

Large customers might have large requirements. If you want to deal with us, they say, you will have to be able to produce one widget every 6 seconds. Such a requirement might need an upgraded plant.

Technological advances (process improvement to improve yields, improve quality, increase efficiency) can create significant profit opportunities. This driver can be part of any of the other categories.

Laws change. Administrations get thrown out and new ones get elected. Government agencies get new directors who have different priorities. Legal requirements (regulatory changes, safety improvements, consent decrees) can change quickly. The government is involved in what comes out of your plant along with the product. Many industries (electric power generation, oil refining, pulp and paper, chemical manufacturing, metal plating, etc.) have had multiple shutdowns to adapt the plant to new air, water, or land pollution rules.

**How do you measure the success of a shutdown?**

One issue in shutdowns is how to measure success. If you were given the job to measure the shutdowns suffered by a series of similar oil drilling rigs, what would you measure? How would you know which one was the best at shutdowns? The measurement of success is essential to modern business. The measures become Key Performance Indicators (KPI) that gauge the effectiveness of the shutdown team and indicate to managers, as the work is done, if things are going as they should.

We would want to look at the performance of the shutdown in key areas that have an impact on the whole company's profit and reputation.

We might start the comparison with the preparation work done before the shutdown. How well was planning done, how effective was the procurement, were the right things done at the right time. Most organizations are good in some areas and weak in others. Basic skills in all seven areas are necessary for productive shutdowns to satisfy your customers, bosses and top management.

## Quick Report Card

| Topic | Grade |
|---|---|
| Organization: How to organize for a shutdown. The best organizations assign a manager who is given responsibility and authority, planners, support personnel, people who have the time, and all the preparation skill sets needed for an adequate period. | |
| Planning & scheduling: Thinking through the jobs, anticipating problems, and developing contingency plans for when something goes wrong. How to perform the work. Defined overall scope, work lists, control, prefabrication, design of schedule, keeping schedule right | |
| Contractors: How to integrate external organizations. Create accurate contractor packages, identify and evaluate contractors, identify sub-contractors, build in carrots and sticks, mobilization plans to insure the right people are there on day one, demobilization plans to manage costs. | |
| Accounting, Costs: How much did the shutdown cost? How to estimate, report, and control costs. How to fund, estimate, refine, develop contingency cost reporting systems in real time, close out. Evaluate financial risks. | |
| Logistics: Organization for the parts, materials, and supplies. Elements of logistics include a site plan, site control, safe routes for lifts, and any off-site management of materials or equipment. Functions are parts receiving, storage, and job site delivery. | |
| Execution: How to manage in the face of reality. Management control, pre-start briefing, daily routines, shutdown of existing plant, work the plan, deal with whatever comes up; keep your eye on the ball until the game is over. Dealing with risks and developing work arounds for risks that were not anticipated. | |
| Reporting: What happened? Were the lessons learned preserved? Are customers satisfied with how they were kept up to date? Can we avoid making the same mistakes again? | |

Once the shutdown is completed there are four criteria for evaluation:

1. Was the shutdown completed on or under budget? Did we hit our numbers? One rig might need more work than another because of variations in the specific oil deposit (unrelated to the maintenance or operations of that rig) so that the pure value of the shutdown is not a good comparison. The variance from the budget is a good comparison. One factor to consider is the amount of work left on the work list

2. Was the plant put back on line on time or early? The cost of the outage might be dwarfed by the amount of the lost revenue for the period. A $10,000,000 outage in a power plant might cost the company $50,000,000 in lost (and unrecoverable) revenue. On the other hand, in some process plants, coming back on line early is a big problem because it takes a good deal of time to start the process.

Operators might not be available, so an early start is useless unless you know about it at the beginning of the shutdown.

3. Did the plant start up smoothly? Sometimes a shutdown takes a week but getting the plant back to stable quality production can take an additional week. One measure of the effectiveness of the shutdown planning and execution is a quick, smooth start-up.

4. Were all jobs completed? In all shutdowns the issue is the work list (otherwise we could do the jobs without a shutdown at a much lower cost). In some industries the work list must be completed before the plant can be put back on line (nuclear power plants or drug manufacturers). If jobs are deferred in these regulated plants there will be questions that must be answered.

5. Were there any lost time accidents? Any unscheduled discharges of material into the air, water or on to the land? This could be the Environmental, Health and Safety measure. Safety is vital in shutdown work because the environment is so dangerous. One measure of all industrial work is safety, and this is acutely true during shutdowns. The goal is no one hurt or killed. By the same token, shutdown work sometimes puts the environment at risk of dangerous chemical spills, and discharges into the atmosphere.

## KPI (Key Performance Indicators)

Additional measures are reviewed in the chapter on reporting, but each of the five measures listed below can be translated into metrics or measures of success. The meaning of the measures will vary from industry to industry but could be used to evaluate plants of the same type.

*Budget:* Budgeted spending to actual spending or budgeted to actual on a job-by-job basis or chart of accounts basis.

*On time:* How many days did the total shutdown last, compared with the number of days scheduled. If the date is important, then how many days early or late is the plant ready to produce.

*Start-up:* There are two ways to look at start-up data. One common one is the ratio of the number of maintenance days for the shutdown to the total days from last good production to first good production. If start-up were important then the number of days from shutdown completion to stable quality production would be a valid measure of success.

*Jobs completed:* You could look at either the total hours completed divided by total scheduled hours, or look at total jobs completed divided by total jobs scheduled.

*Safety/ Environment:* The usual safety measure is lost time due to safety per million hours. If the shutdowns are similar you could just use the number of incidents per shutdown as a simple metric. If the events were of different sizes, then dividing the number of incidents into the number of hours in the shutdown would make the numbers comparable. For an environmental metric look at the number and type of incidents.

## What is the basic difference between Traditional Project Management and what we do with a typical shutdown?

There are major differences between project management as a field that you can study or take courses in and what is done in a maintenance shutdown. Project management is an area where there are courses, training materials, and many people with expertise. They might be programmer project managers, construction project manager or military program project managers. All these jobs are sufficiently different from shutdown project management that the skill sets are not always comparable.

There is overlap. Projects such as building a new assembly line might be part of a wider shutdown. Elements of the shutdown are more like traditional projects.

Both projects and shutdowns are temporary events but:

| Typical Project | Maintenance related shutdown |
|---|---|
| ▪ Many related jobs | ▪ Many unrelated jobs (1 per work order) |
| ▪ Logical steps interrelated to an end result | ▪ More unknowns and greater emergency work. |
| ▪ Scope of work is usually pretty clear and does not change | ▪ Many one-step activities |
| ▪ The work is usually visible | ▪ Scope of work is not always clear and is undefined until right up to the beginning of the shutdown |
| ▪ Project is organized around cost codes, and a hierarchical job structure | ▪ Scope will change as items are disassembled. |
| ▪ Schedule can be updated on a weekly or even monthly basis | ▪ Much of the work is invisible inside tanks, rebuilding pumps etc. |
| ▪ Understandable end point (building is complete) | ▪ Planning must wait until the scope is pinned down,which may be late in the process |
| ▪ Less need for safety permits and clearances | ▪ Updating must be by shift or even more often |
| ▪ Can be planned well in advance | ▪ Extensive safety permitting is needed for every shift |

| Typical Project continued | Maintenance related shutdown continued |
|---|---|
| ▪ Different use of software; a big issue is critical path analysis. | ▪ Staffing levels can vary wildly |
| | ▪ No end point (there's always more work) |
| ▪ Staffing requirements are static | ▪ Big issue is resource leveling |
| ▪ Span of time is days, weeks, or months | ▪ Span of time measured in hours and shifts |

In military parlance a related group of projects is called a program. So the Polaris missile (where the PERT system was developed) was a program of dependent and coordinated projects.

## Institute a long-term shutdown strategy

The most effective approach to maintenance shutdowns, outages, and turn-arounds is long term. The plant top management should look a decade into the future and decide in a broad brush way what will be shut down and when. The top brass of the plant or group would review the long-term strategic plan after every major shutdown to see if it needs to be modified.

The timing, budget, and scope of individual shutdowns would be managed from shorter-term tactical plans. The tactical plans would include exact timing based on specific long-term spare lead-times, commodity pricing, and market and business conditions.

A power generator in the Caribbean maintains a 5-year plan for shutdowns in each of its LNG power stations. As management plans and executes the individual shutdowns they update the long-term plan. Using a plan like this they can predict shutdown budgets and requirements well into the future. A pharmaceutical company has a 10-year shutdown plan. Within that plan each train (connected series of processes to make a product) within the plant has a major shutdown every other year and a minor shutdown on the off years. Management is always looking at what will be in the next major and next minor events.

The long-term approach allows the most important choice of all. That is the choice to get training and experience so as to be 'great' at shutdowns. The company can make the decision to build muscle in this area.

**Since this is a long-term part of our business, let's build expertise in shutdowns as a core competence.**

**Is your firm on this path?**

**Are you part of this effort?**

# CHAPTER 2

# SHUTDOWN COMMUNICATIONS

Shutdowns are big events. As in all big events the stakes are high and there is significant anxiety from people at all levels in the organization. In addition to anxiety there is a significant amount of specific information held by one person that is needed by another. The key to successful shutdowns is constant communications.

## If no one reads your E-mail can you say communication took place?

Communications is a two-person process. It is not enough to send out a fifteen page E-mail and expect that communication has occurred. Communication requires a transmitter and a receiver (or a transmission and a reception). If the recipient doesn't read the E-mail then no communication takes place. There is a responsibility on the part of the shutdown management team to use techniques to ensure that communication takes place and that people read their E-mail relating to the shutdown, and think about it.

To encourage communications we create an atmosphere where people can freely reveal problems. If people are concerned about consequences and sanctions when they report problems, the atmosphere will not be conducive to solving the problems. Everyone has heard the stories about the king who killed the messenger who brought bad news. In shutdown teams, messengers who bring bad news need to be praised and lauded and not killed! Only when it is on the table can bad news be managed.

John Larson, a ChevronTexaco Project manager working on a project-driven (as opposed to a maintenance-driven) shutdown offshore in the gulf of Thailand, has one rule. His rule is "no surprises." He is scrupulous about making sure his boss and the customer know any bad news that he is aware of and he requires that his team keep him appraised of any adverse situations. This requirement doesn't mean that he gives up the ability to solve problems. On the contrary, he might give his boss bad news with an action plan to mitigate or avoid the problem.

## Basic Skill: Meetings

The dictionary says that the definition of meeting is to come face-to-face or to come together for a common purpose. In shutdowns the face-to-face part is less important than the common purpose. Some key people can use technology to avoid being face-to-face to attend a meeting by phone, videoconference or over the Internet.

Ormen says, "Project status meetings are both communication and control devices. This is the forum where people keep up on what others are doing; where you build the cornerstones of your project culture and set the tone for your people (from the Digital Norseman's, PM web site. Details can be found in the resource section).

Every shutdown is managed by a series of meetings. Even with all the improvements in technology, meetings are still the key communication tools for shutdowns. Improving your organization's meetings is a high return on investment activity. A small improvement in meeting efficiency could be the single most important thing you can do to improve your shutdowns!

Shutdown meetings loom large at the beginning of the process and are designed to get people up to speed, solicit input, and make sure nothing is forgotten. As the shutdown progresses the meetings get smaller and are more focused on particular issues.

How many of you had a class in how to run meetings? Is there even a course in this important skill? This essential skill should be taught on day one in any school that wants to train people for business.

In general it is better to have two short meetings per week rather than one long one. Even long shutdowns need frequent updates and communications. Shorter shutdowns will have daily shutdown meetings. Usually meetings start with a high energy and with intention (to solve a problem, communicate a plan, etc.). After an hour people lose their focus and their intention starts to erode unless the meeting is managed very closely. Sometimes a quick 15-minute stand-up meeting is all that is needed.

The key to keeping projects on track is to detect problems early enough in the process that an intervention can be made before it is too late! It is no good finding out at a weekly meeting that an eight-day shutdown is only half complete.

As we'll explore in the project management section, one of the biggest issues of shutdowns is choosing meaningful activities. If your activities are well chosen, then progress tracking is not difficult. Consider creating a watch list of deliverables to be reviewed at the daily shutdown meeting of all activities due within the week on or near the critical path. Make sure you keep getting feedback on the status of these activities in addition to your regular agenda items.

Meetings are important but it also won't hurt to get out on the jobsite daily and see for yourself what is the status of these critical and near critical activities.

Of course, if you go out into the field you'll hear about small problems that won't make it to the meetings. In person this process is quite messy, but as they say in the army "no battle plan ever survived contact with the enemy."

In addition to meetings with the project team you need to keep the stakeholders and your boss updated by means of less frequent meetings (monthly) with a broader overview of the project. In complex or longer shutdowns (installation of a cogeneration facility, for example), these meetings serve to re-engage the stakeholders and the bosses in the project (especially during the dark days).

The following are the most important areas of meeting management: (you might focus on one at a time until you feel you've gained some mastery in that area).

*Agenda:* The agenda is the road map for the meeting. Without it the meeting will meander (but you might not notice it since there is no agenda).

*Attendees:* Who is in the room is almost as important as what is discussed. Be ruthless about who you invite.

*Notice:* The best way to get the right people to attend is to make sure they know about the meeting. Be rigorous.

*Practices:* These are things to do to make your meetings work. As with all skills you have to repeat these practices before they will feel natural (psychologists say repeat the new behavior up to 21 times to feel natural).

*Accountability:* Holding people accountable for what they promise or what you've requested is important to the smooth functioning of the whole shutdown. People tend to avoid accountability. Once they get a taste of accountability without blame they will realize that their growth and development is linked to the accountability that they take on.

In *Turnaround Management* Tom Lenahan lists some of the

meetings necessary for getting set up for a shutdown as follows: (other meetings may be needed to actually run the shutdown after it commences).

1. Policy Team Meeting
2. General Work list Meeting
3. Major Task Review Meeting
4. Project Review Meeting
5. Inspection Review Meeting
6. Spares Review Meeting
7. Plant Shutdown and Start-up meeting
8. Safety Meeting
9. Quality Meeting
10. Site Logistics Meeting

To make them most effective these meetings have some additional guidelines:

- The agenda is restricted to one topic based on the title of the meeting
- Only those who are directly involved are invited
- Clearly define the block of work or time limit and do not exceed it
- The chairperson should be the most appropriate member of the Preparation team (the logistics meeting, for example, might be chaired by the logistics manager)
- Minutes are prepared following the advice above for general meetings and copies are sent to the shutdown manager. Copies are also included in the shutdown narrative.

| **Topic** | **The 'good meetings' check sheet** |
|---|---|
| Agenda | ❑ Have agendas passed out in advance and in a form that is understood by everyone. How do you know if everyone understands it? Try asking privately. |
| Attendees | ❑ Decide who should be invited and determine if a particular person is needed, or if a representative from that area is needed (and communicate that fact). The person's own good time management techniques may have someone tied up for your meeting time so if you can do with another representative, that's great. |
| Attendees | ❑ For scheduled meetings, periodically review the invitee list and see if everyone should still be invited. Talk to anyone you are dropping. Add people as the shutdown reaches a stage where their skills are needed, and drop people whose skills or contributions are no longer required. You can ask. |
| Agenda | ❑ Consider putting the most important agenda items second and third on the agenda (not first, to accommodate late-comers). |

| | |
|---|---|
| Agenda | ❑ Consider putting times on agenda items. (Discuss problems and decide on a course of action such as "contractor ABC 15 minutes…") Some leaders only put the time on their agenda to guide them. Others use time publicly to help discipline the whole group's discussion process. |
| Agenda | ❑ Tailor the number of agenda items to the type and length of the meeting. |
| Agenda | ❑ Consider reviewing the agenda with your boss ahead of the meeting. This would be particularly useful if there are any sensitive or controversial agenda items. Remember the commandment "Thou shall not surprise your boss on sensitive issues" |
| Notice | ❑ Give as much notice as possible for ad hoc meetings (as opposed to regularly scheduled meetings). |
| Notice | ❑ Publish and publicize the start and end times clearly. If someone is late it should never be because they were not informed of the time. By the same token, when a meeting time has been changed, send the change out and request a reply acknowledging the change. Call or page people who do not respond. |
| Practice | ❑ Start and end on time and expect people to attend on time. If they are late expect an apology but not a reason or excuse. If a person is late (especially without communicating with the leader) they have just wasted everyone's time. It doesn't really matter why. |
| Attendees | ❑ Know who will not attend. Have people know they should contact you when they can't make the meeting. Everyone's time is important, if you know that someone cannot make the meeting you'll know that you don't have to wait for him or her. |
| Attendees | ❑ In regular meetings where the same people attend put out nametags (called tent cards), which will remind people of everyone's name. The nametag at the table (without a person) also shows who is missing. |
| Practice | ❑ Ask everyone to set his or her pagers and phones to vibrate (or if they can risk it, to turn them off). Calls should be taken outside the room unless the content directly impacts a discussion or decision. |
| Practice | ❑ As the leader, be the first speaker and quickly sum up the status of the shutdown then provide a short overview of the agenda of the meeting. |
| Practice | ❑ Report on deliverables and progress (or ask for reports from the people who are responsible for each area). If appropriate, reprint the Gantt chart showing progress areas and problem areas |
| Accountability | ❑ Assign tasks and responsibilities, being sure to include in the minutes who is to carry what out by when. Of course how and why are useful but not essential. |
| Accountability | ❑ Follow up on tasks and responsibilities assigned at the previous meeting. |
| Accountability | ❑ Insist that meeting goers do their homework |
| Practice | ❑ Take straw polls on controversial issues ahead of time |
| Practice | ❑ Review important presentations ahead of time – do not be surprised. |
| Practice | ❑ Have someone take minutes (be sure to select someone who is good at it). |

| | |
|---|---|
| Practice | ❑ Publish brief minutes that include promises made, deadlines, and follow-ups. Have a simple system to manage promises and requests. Copies of the minutes get filed into the shutdown narrative report. |
| Practice | ❑ It is important that when decisions are made they are formally written into the minutes and all attendees agree to support them even if they did not agree with them during the discussion. |
| Practice | ❑ Bring issues forward from one meeting to the next until they are resolved. If any item is important, make sure it is resolved. As mentioned it is important to have a full discussion where all views are aired. When a decision is made the team should stand behind it (even if individuals don't agree). |
| Practice | ❑ Meeting length is proportional to the comfort of the room (no chairs means a short meeting) and to the level of refreshments (add food - add an hour) |
| Practice | ❑ Chose an appropriate place for the meeting and manage temperature, sound level, light level, and distractions if possible. |
| Practice | ❑ Encourage the practice of looking for failures in your system rather than looking for people to blame. |
| Practice | ❑ Create an environment where people reveal what is going wrong first when they speak and what they plan to do about it. They can also ask for help. |
| Practice | ❑ Have a practice of honest acknowledgement for accomplishments |

# CHAPTER 3

# SHUTDOWN PHASES

The techniques of Project Management (another use of the acronym PM) can be used to model the maintenance shutdown and some of the techniques can be used to control the project. There is a good deal of training available and a very accessible body of knowledge about project management.

This body of knowledge (in the field this is referred to as the PM BOK or Project Management Body of Knowledge) consists of techniques and strategies to efficiently manage large projects. Much of this PMBOK applies to maintenance and project driven shutdowns.

The shutdown planner should be conversant with the major PM techniques and the ways to set-up and manage jobs. As mentioned, training is available in PM techniques in almost every major city and at most universities. There is a threshold where a very large maintenance job would benefit from Project management techniques (and one where a short, small, shutdown could just as well be done manually without the PM techniques).

To use these techniques effectively, project managers divide the shutdown into phases. Completion of each phase is marked by the completion of a deliverable. By setting up phase-ending milestones as deliverables, managers outside the project can review progress without getting mired down in the details and insure compliance with budgets and business objectives.

In a maintenance shut-down or major maintenance project the phases might be:

**Phase 1:** The job is initiated. This phase answers the questions what and why? Top management is deeply involved at this point. They set the context within which everyone else will operate. A preliminary backlog is run for review and to secure

| Phase | Description of Phase | Deliverable |
|---|---|---|
| Phase 1 | Feasibility, initial discussions, initial scope of work and budget, approval of approach. Also called Initiation. | Completed feasibility study. Documentation in hand for engineering to begin, business case study complete |
| Phase 2 | Plan, Design and Engineering phase includes engineering, drawings, planning documentation, and firm budgets. | Planning package, Completed Job list, engineering drawing package and bidding packages available for review. Contractor is chosen. |
| Phase 3 | Production and execution process includes building, assembly, removal and replacement. | Ends with substantial completion of shutdown job list. Delivery of punch list items. |
| Phase 4 | Turnover and work completion is completed with quality assurance, life safety testing, and start-up completed. | All shutdown work completed, accepted by end-user, start-up completed. Plant or facility in normal operation. All bills from vendors and contractors booked. |
| Phase 5 | Completion of Project paperwork and closing out and review of meetings. Also the beginning of a new Phase 1! | Final manuals, as-built drawings, updated planning packages, operator directions or manuals. Lessons learned book written and distributed. Index created for files. Final job list of all jobs not completed in this shutdown ready for the next one. |

some agreement of the most important jobs to be done.

**Phase 2:** Also called Preparation. All planning is completed. Organization for the shutdown is defined. All stakeholders have had input. The question that this phase answers is how?

**Phase 3:** Work execution begins. Tactical expertise (and occasional heroism) keeps the majority of jobs moving toward completion. This work group is the largest to be involved, for the most intense and shortest period. Three parts are shutting the facility down, doing the work, and bringing the plant back into operation.

**Phase 4:** All traces of the shutdown have been removed. This is also called Termination. Any punch list items have been completed or have been accepted incomplete. The customer is happy and the results are in line with what was requested.

**Phase 5:** At this point review all files, create indexes, and decide what to keep and what to discard. Write up a narrative for the lessons learned. This is also called the Interim stage (waiting for the next project).

## It's all in the details

To effectively manage a shutdown you must manage at a level of detail that will make a difference in the outcome of the project. In a nutshell, when you apply Project Management to a shutdown you list all the activities to be performed; what resources are needed, and which activities are dependent, interdependent and independent.

Since the amount of detail can be enormous, the Shutdown Manager will assign the details to the person(s) most likely to understand and be equipped to deal with it (this is frequently a contractor). The manager and the team build management structures that remind them when to ask questions and what answers would make a negative or positive impact on the project.

To remind the manager of the important items to manage each day (or each hour!) most large projects are loaded into software. From all the information loaded, the software builds a model of the project. This model should represent all the activities for all the projects that make up the shutdown, along with their critical elements such as durations, effort levels, and other resources.

Under the best conditions, as the various activities are completed the shutdown management team updates the model. When properly fed and watered, the PM software tracks the shutdown real-time and provides alerts when activities fall behind or when there are resource conflicts. Computer software has proliferated and thrived in this area.

## Timing

Each of the five phases has a specific timing. If they are displaced from their timing they will not serve the purpose for which they are intended. For example, conducting the planning process in the middle of the job while the steel is going up is not as effective as planning before everything starts (although there is some advantage even then).

The five phases can be mapped onto PMBOK's (see resource section) process groups.

Another way to look at these phases is to regard a shutdown as a series of interweaving processes (from PMBOK guide published by Project Management Institute). You can see that the phases on the project overlap,on the next page.

The only process without an analogue in the phases diagram is the controlling process. In shutdowns, each phase has its own controlling process, which can be exercised to measure the

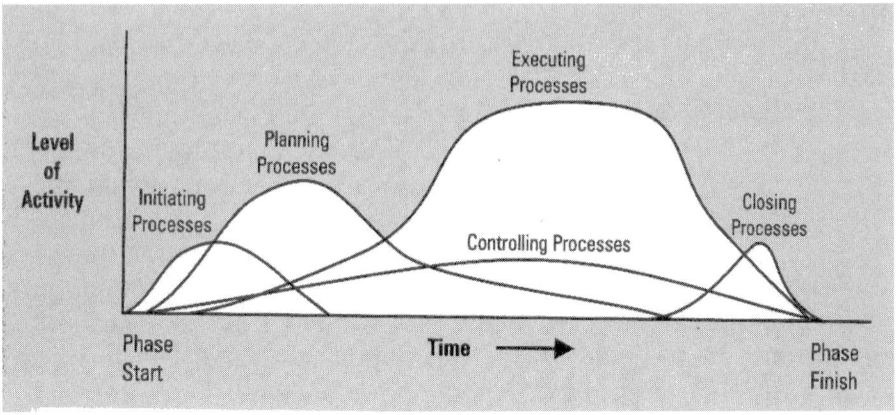


effectiveness of that phase. In the phase model there is an additional phase connecting the finish to the start of the next shutdown. Unlike discrete projects that the PMBOK deals with, shutdowns come at you like waves at the beach. There is always one behind the one that is coming at you. Each wave builds on the energy and action of the waves before it.

### How do you currently manage shutdowns?

The best management style will fit the needs of the organization and the project. The best style depends how well the project is planned, how well the teams know the work and each other, and how clever the organization is at project management. If you hire a specialist to change the catalyst in your refinery and add in a batch of work you will utilize a very different style than you would if you attempted to do a complex job yourself for the first time.

**Reporting style:** The reporting style is a project management office that only reports on what happened. This kind of project manager will report what happened and what is likely to happen, given what has happened and what needs to happen. The action is left up to others. The process is called the news reporter style where the anchor reads the news but isn't supposed to add or subtract anything.

**Control from above style:** In the control from above style, the project manager stands clear of the action and in addition to reporting, issues commands and moves troops to work in problem areas. These people manage from start to completion times of individual activities, but leave the details to the field. The style is like that of the coach in a football game. The controllers call the plays and watch the results but do not jump in to join the action.

**Control from the center of the action:** This control is a hands-on, get your hands dirty, project manager. You'll see the manager in the thick of the project giving advice, yelling, begging, threatening (contractors, workers and anyone else who gets too close), to keep the critical jobs on track. Think about them like a fire fighter crew chief, who jumps in where necessary, when necessary.

**Shutdown Cycle**

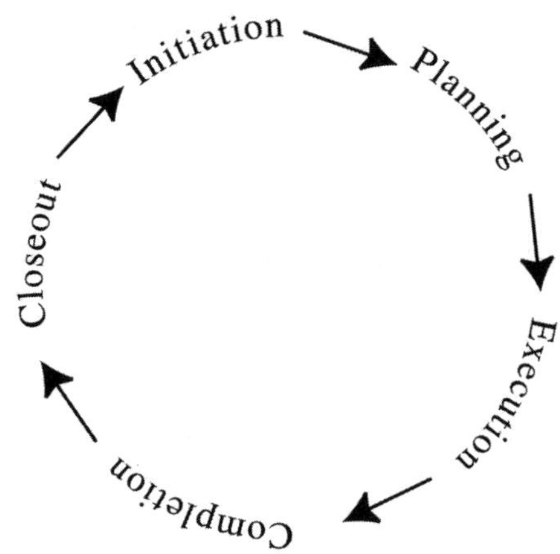


If you think about shutdown as a process like the seasons of the year you would not be far off. Although this shutdown seems like an isolated, unique event, in fact it is part of an ongoing process.

# CHAPTER 4

# How to Justify the Shutdown

Justify the project by building a strong, documented business case based on what is the urgent need. It is wrong to assume that work has to be done at any cost. The case must be made on a business basis. You are presenting the facts for your shutdown to what Tom Lenahan calls the Policy Team.

**Policy Team (AKA Shutdown Steering Committee)**

Because of the size of the shutdown (usually large when compared to the size of other maintenance events at that plant), you want top management as well as major stakeholder groups to be represented. This team includes business, engineering, turnaround, and even marketing managers. To be a member of this policy team the members should either be directly affected by the outcome of the shutdown (such as operations or marketing), be the provider of funds for the shutdown (top management), or have authority to make decisions about the shutdown (shutdown manager)

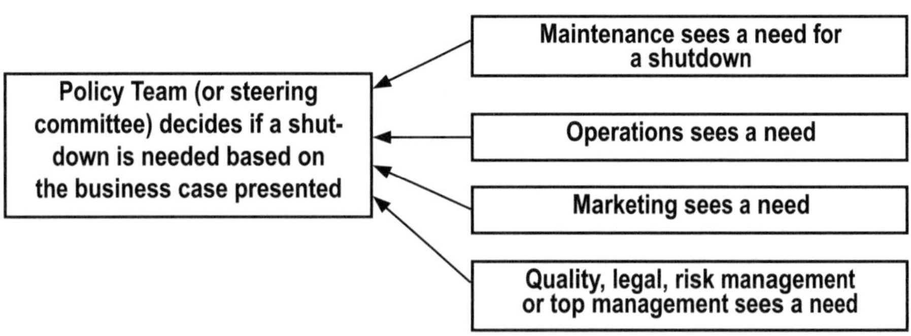

The Policy Team provides funds, establishes constraints, sets objectives, establishes policies, and monitors progress of the planning, execution, and completion of the shutdown. This policy team stays in existence from the early planning throughout the entire shutdown and may dissolve after the shutdown close out process is completed. If shutdowns are an ongoing part of the business, the policy team will stay together (recommended).

They hold the shutdown manager accountable for the performance of the shutdown. The turnaround manager will regularly update the policy team with milestone charts, reports, budget variance reports, and forecasts. Approval for additional funds due to added jobs would be agenda items. Sometimes the Policy Team serves as the bad guy and rejects added work and pet projects, thus allowing the shutdown manager to be the good guy.

### Is this shutdown needed at all?

First question to answer: Is this shutdown needed at all? Is there another way to provide the same outcome? Can we rent the capability while we work on our own? For example, work on a substation might be possible without a lengthy shutdown if a substation could be rented and used to bypass the shutdown unit.

### All shutdowns operate within constraints

The biggest constraint is the amount of money you are given and the duration. Other constraints might include a business cycle window, work force, availability of contractors and workers, the complexity and hazard presented by the plant, or even the profitability of the industry.

### Important distinction: Shutdown days and maintenance days

*Shutdown days* are the number of days from the last full good production to the first full good production after the shutdown.

*Maintenance days* are the number of days the shutdown maintenance and project work goes on. Days for cool down, decontamination, and startup are subtracted to yield maintenance days.

A constraint can be as subtle as a few paragraphs in a bargaining agreement. In a small utility the agreement was very restrictive about the use of contractors on the site. As a result they used few contractors, which prolonged the shutdowns. A 30-day outage in another company might go 45 days in this one.

In the pulp and paper industry the cost of the raw material varied over the years. When the cost was high, the plant ran 24/7 and made as much money as they could. When the price went down the plant was slow. So far so good. The maintenance department budget was based on a percentage of sales. The problem was that the shutdowns couldn't be properly funded because; when they had the money from sales they didn't have the time. When they had the time, they didn't have the money. If only they could be allowed to stash away some money from year to year for the inevitable rainy day.

## Minimum Contents of the justification package for management

Each organization has rules about what is contained in the scope of work to be submitted to management. The justification should be produced at the level of typical reports and requests for funds in the company. If requests from other departments to management are delivered on glossy paper with 4-color charts, pictures, and artists' renderings, then yours should too. At the very least the document should include:

- An executive summary on one page or less
- An outage justification statement
- A description of the major work to be accomplished (pictures are useful)
- Reasons why the work has to be done at all, and why a shutdownis necessary
- Why the work has to be done now.
- The major deliverables for the project with individual justifications
- A sampling of minor deliverables (that are know at that point)
- Proposed duration of shutdown, known risks to that duration
- Proposed budget for shutdown by major project, risks to the budget
- ROI (Return on Investment) calculations if there is improved efficiency or effectiveness.
- Other possibilities could include constraints, assumptions, and historical information.
- Major risks and contingencies involved

**Interesting Analysis:** How many days of production are needed to pay for this shutdown? (Calculate this figure by dividing the total cost of the shutdown plus the total cost of the lost revenue for that plant by the daily profit of the plant when operating).

## Marketing a maintenance project is aimed at 'Spinning' the story to what the audience is vitally interested in

Maintenance is viewed as dull. When maintenance needs to replace a boiler it does not add anything fun or sexy to the company (like a new product or a new customer). Maintenance projects merely maintain the status quo, like maintaining efficiency, or obtaining a few extra BTU of steam. Maintenance projects may have to be gussied up to appeal to various stakeholder groups in the facility.

Thus, when the project is generated by maintenance it must be sold to other groups in the organization. You'll hear if it ain't broke don't fix it and other such nonsense. This shutdown must be made to appeal to the self-interests of each kind of stakeholder. Consider discussing improved maintenance investments using the issues shown in the following table.

**A Bakers Dozen Real Benefits you can use in your next justification**

| Shutdown Feature | Benefit |
|---|---|
| 1. Reduce cost/unit (cost per ton of coal, cost per car, cost per container unloaded). | More profit |
| 2. Reduce downtime and increase plant reliability | More profit |
| 3. Increase quality of output with new processes | More profit |
| 4. Reduce the number, size, and scale of repairs (and even the need for repairs) using new technology or equipment | Less headaches |
| 5. Increase equipment availability from new or refurbished components | Conserve cash |
| 6. Increase remaining useful life of equipment | Conserve cash, More profit |
| 7. Comply with the law and avoid fines (and jail time!) | Keep Out of Jail, keep out of newspapers |
| 8. Decrease potential exposure to liability (improve control of by-products and processes) | Keep out of court, keep out of newspapers |
| 9. Increase operator, maintenance mechanic, and public safety | Keep out of court, keep out of newspapers |
| 10. Reduce future maintenance costs | More Profit |
| 11. Reduce overtime due to fewer emergency breakdowns | More Profit |

| | |
|---|---|
| 12. Reduce investment by not needing spare or stand-by units (put in one 'big one') | Conserve cash |
| 13. Get an opportunity for permanent solutions to problems that have stymied people for a while | Less headaches |

### Why is this list of benefits important?

At least one of these features and the resultant benefits should be of interest to every stakeholder. Of course, different ones might be interesting to different stakeholders. To determine which benefits would appeal to which stakeholders you need to know the stakeholders well. Do you know the stakeholders well enough to assess their priorities?

Consider how the interests would change if you were talking to maintenance workers, accountants, or purchasing agents? Every type of employee, in every role in your firm, has a point of view and interests that they hold dear. When you propose a change to the way business is conducted, consider these points of view. Identify your organization's (and its individuals ) priorities and speak to those priorities

### What is the bottom line for shutdowns in your organization?

One bottom line is that the plant had better be more reliable after the shutdown than before it.  Although reliability is usually increased, sometimes the plant is actually less reliable afterwards!

### What is the Shutdown covering up?

One fascinating question is why do we need a shutdown? Often the reason is that the shutdown will cover up something that, consciously or unconsciously, the managers would rather not have to face.

In *Turnaround Management,* Tom Lenahan makes a good point about organizations using their shutdown cycle to cover up bad practices. These problems come in several flavors including covering up bad policies and procedures, bad designs, unsuitable building and installation, or bad expectations.

Some shutdowns may have been caused by bad operation of the plant. How many managers want to face that fact? How many will fund retraining, or replacing the leadership. Perhaps the incompetence is related to top management attitudes, or the unrealistic expectations of management are driving the inadequacy. On the other hand perhaps the regular maintenance crew is incompetent and the shutdown is aimed at periodical clean up of the mess.

An example of bad procedures would include fixing chronic problems over and over rather than investigating the source or root cause and fixing that. Some organizations have been doing shutdowns for decades. Items appear on the shutdown list as a result of custom or traditions. How many traditional activities find their way on to your work list (items for which no one knows the reason, and everyone is afraid to remove them)? Or, in another popular area, what about cover-ups related to inadequate attention to PM procedures and processes?

In the bad design arena many shutdowns are used to cover up an inadequately designed plant, or a plant with inadequate environmental protection. What about shutdowns arising from inadequate commissioning, installation, or construction? Do you have any of these?

Investigation into these areas can be fruitful because you might get the opportunity to fix some of the problems permanently and avoid or delay the next shutdown.

## Do you have a sponsor?

Sponsors are people in high positions who will back your shutdown in meetings and private conversations with other top managers. They could be ex-maintenance or operations people who have moved up to the executive suite. It's always good to have a sponsor. Some projects are controversial and the problems might transcend project management and get hung up on political infighting. Still, the sponsor can only open the door. You will have the pleasure of getting the whole team through it!

## Keeping the stakeholders in the communications loop

- Identify stakeholders
- Update and provide consultation at milestones concerning project status
- Update and provide consultation about progress at the end of each phase

One of the most important early tasks is an identification of the stakeholders of that particular shutdown. This is a non-trivial task because a shutdown has many different tasks that can have different stakeholders. The stakeholders are all the people who will be impacted by the results of the project. Their involvement is essential for the company to provide wide perception of the success of the shutdown and even the success of the whole plant, process, or capability. Some aspects of the importance of stakeholders and their stakes are shown in the following table:

| Stakeholder | What they have at stake |
|---|---|
| **Top Management** | When his/her bonus is based on profitability, Shutdowns represent high risk/reward opportunities. Wants to avoid headaches, or anything that could look bad to board, might need a success in the shutdown to look leader-like, or to be considered for a promotion. |
| **Production manager** | Total performance tied to quality production, shutdown could also be a major headache or a big opportunity (high risk-high gain), could increase output and improve quality, could reduce current headaches by solving problems long-term, could open a giant can of worms. |
| **Accounting manager** | Shutdowns are a major pain in the neck because they disrupt orderly work, could consume excessive hours chasing missing documents, are high stakes for auditors who to want to review how shutdown is charged to books, and shutdowns will always make more work. Also shutdowns are on the cutting edge of important decisions about capitalization, expenses, profit, and depreciation. |
| **Maintenance manager** | Chance to shine or crash and burn, time of uncharacteristically high visibility, show quality of team, may be tied to bonus or review, miss family time |
| **End customer** | Wants high quality, steady, reliable, and cheap production, can make buyer look bad (or if successful look good), might make buyer look elsewhere if the shutdown goes badly |

Good business practice says these stakeholders should be consulted at the end of each of the major phases. Do you remember ever being on the other side of a project or shutdown? Do you remember getting stuck in an unworkable situation that was unworkable because you were never consulted? In too many projects (such as a new process or building) not all of the stakeholders were able to have their say. Their wisdom was not available to the project planners and the organization is stuck with an inadequate asset, machine, or building.

### Presentation Trick

One trick to gain support is to get the key decision-makers or stakeholders on the user side to participate actively in making the estimates, wording the justifications, and giving the presentation.

If possible, a team must be chosen that will span those stakeholders who will be impacted by the change. By involving these people early in the process we can help assure large changes in scope are considered as the project is developed rather than after it is approved.

# CHAPTER 5
# TIMELINES AND AUDITS

## Prepare the Timeline

Take the list and add completion dates to major planning events. Defining, planning, and creating a schedule is a project itself. A large chemical process plant hires a contractor to work with them to define, plan, and package the jobs. They have found from experience that the amount paid for that service is dwarfed by the amount saved.

Even planning a shutdown has a timeline and critical path. Certain planning elements had better be in place on schedule or the shutdown will be delayed. Or, worse than a delay, the planning will be incomplete and the shutdown will not be properly planned.

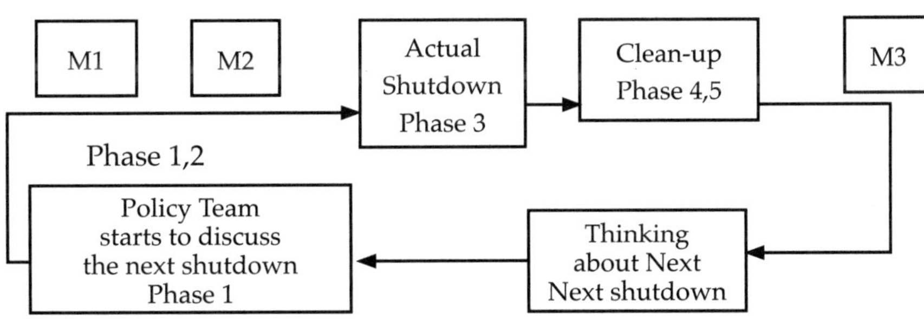


M1 Initiation milestone and audit poin
M2 Planning milestone and audit point
M3 Completion milestone and audit point

If you take the long view, the phases of the shutdowns are like the seasons of the year. Each season leads to another. Winter might be over for now but there is always another coming down the pike. If we are careful and put away our winter things properly, they will be ready next winter. If we are haphazard we will not get any advantage from the lessons we learned last winter. Humorously (or perhaps not) some firms act as if this is the first and last shutdown ever.

The best practice is always to be thinking about improving the shutdown and devise a system to keep track of these good ideas.

**Potential Milestone points prior to a moderate-sized shutdown**

| | |
|---|---|
| Initiation and project definition | -18 to36 months |
| Order very long lead time spares | -18 to 36 months |
| Budget approved | -18 to 36 months |
| Initial Project meetings | -10 to18 months |
| Meetings with stakeholders to gather jobs | -12 to 16 months |
| Meetings with Turnaround Team | -6 to 9 months |
| Engineering completed | -8 to 12 months |
| Contract document started | -9 to 12 months |
| Contract documents completed | -7 months |
| Contract awarded | -6 months |
| Organize pre-shutdown work | -6 months |
| Initial PERT/CPM charts worked | -6 months |
| Work list close out date | -6 weeks |
| All work validated (validation has been on-going) | -6 weeks |
| Logistics ready to go, approved site plan posted | -6 weeks |
| Initial plan generated, final network charts | -5 weeks |
| Delivery of substantially all materials and parts | -4 weeks |
| Final all-hands Briefing | -1 week |

## Scheduled Audits

A skilled team is chosen to audit shutdown preparedness in all plants of an organization. The team should consist of volunteers from other plants. If you only have one plant then appoint a team not involved in the current shutdown but who have some experience with shutdowns in general (even if you have to go outside the company). Retirees, contractors and consultants can be recruited for this job.

The team (in its best form) audits the shutdown process at several points (M1, M2, and M3 above). Where deficiencies are found the team comes back in a short time and conducts training and individual coaching as needed.

The goal is that company wide, all shutdowns take advantage of the lessons learned from previous shutdowns and continue to improve.

The audit team (it would be great if shutdown team members from other plants would volunteer for shutdown audit teams) covers things like:

1. Is the approach and strategy adequate to produce the desired result? This question would be settled during the first audit (M1).
2. Are the systems in place, and the system that will be put into place, adequate to manage a shutdown of the size contemplated? These systems would include (but not be limited to) planning, procurement, engineering, contracting, safety, security, accounting, work management, accounting, and archiving (M2).
3. Advance planning milestones (are enough jobs planned at each milestone)
4. Engineering items such as drawings and specifications
5. Are contractors in place (in time) and are they qualified for the work?
6. Is the budget complete
7. Are PM and PdM inspections complete, and are deficiencies identified
8. Is staffing worked out?
9. Are there 'real' organization charts prepared for the shutdown?
10. Is the amount of staffing in the shutdown core group adequate and is the amount of time for preplanning adequate? How far in advance must the milestones occur to make the project work?
11. A few weeks before D-Day, is the shutdown ready to go?
12. During the shutdown, is the shutdown team in control of the event? Also are methods, safety, and environmental compliance, in line with policies.
13. Is the shutdown closeout complete (M3)?

Each organization should create a series of metrics that are benchmarked for each audited plant that can be tracked over time.

## KPIs

KPIs (key performance indicators) for this shutdown were designed at the outset of the shutdown. The goals were shared with the entire team including the contractor. The shutdown team has been recording KPIs in the five areas. The final KPIs are recorded in the shutdown report. The audit group reviews these measures.

# CHAPTER 6

# THE SHUTDOWN ORGANIZATION

The team is the key to success of the shutdown. Clear communications are essential to ensure that the thousands of details are handled at all (let alone in the best way). In shutdowns and battles there are really two groups; the group that plans the battles and the group that fights them.

Before you assign the teams consider the important question, what skills sets are needed for the job? Too often, people are assigned to shutdown teams because there is nowhere else to put them or they happen to be available at that time rather than because they are the best for the job. Take a cold look at the people and decide if you are not better off bringing in an experienced outsider and reducing the number of staff people involved.

If the shutdown is some time away you might consider investing in training or lending a person to a sister plant to gain experience by assisting in their shutdown. Many industries have special interest groups devoted to improving shutdowns (electric power generation, oil refining). Even if you are not in one of these industries and your industry does not have meetings on shutdowns you may be able to attend meetings of such a group.

This is also a great opportunity for double-teaming positions where older, experienced persons guide a younger person and hopefully pass on some of their knowledge and experience. Some organizations give small shutdowns to newer people. Training on small shutdowns compresses all the aspects of larger events into a manageable time frame. Learning can come fast and furiously.

Any member of the team (including the manager) can be an employee, a consultant who is an expert in shutdowns, or their area of the shutdowns, or an employee of the main contractor.

## Shutdown Project Team

- The organizational structure is a written document that will be widely distributed
- It is assumed that the people have the skills for their roles or are being trained and overseen by people who do have the skills (double teaming).
- The planning team lays out the shutdown and prepares the shutdown packages. The team has a strategic view and always looks for ways to avoid potential problems.
- Once the shutdown starts the action-oriented execution team takes over. The executioners have a tactical view so as to keep all the balls in the air until the project is completed. The personnel can be the same for the two activities but their roles shift from strategic to tactical. Keep in mind that the personality of a great executer is different from that of a great planner. The simile is about the same as that between battle planners and battlefield commanders.
- The team has to manage the human issues of the shutdown such as providing leadership, communicating, delegating, motivating, coaching, resolving conflicts, working on labor relations, etc. A shutdown is a melting pot in accelerated time, which means that people will be operating at or near (or over) their limits.

**One common question:** Can the shutdown team also have regular responsibilities and do their job part-time? This question is a tough call and depends on the size and complexity of the shutdown and experience of the personnel. Different models are project driven, matrix driven, and function driven. Any type of model can work on smaller events. Complex, large, and critical shutdowns usually require a project-driven core group (at least).

The lines on the organizational chart are lines of decision-making and authority. They (should) answer the questions; who has the right to make the decision, who is responsible for an action (getting it done), and who is accountable.

## Shutdown organization should insure that:

- All function are covered by someone
- There is enough depth to be sure that more than 1 or 2 people understand the technical issues of the shutdown

| | |
|---|---|
| **Project driven** | People are members of the Shutdown team completely and they report to the Shutdown Manager. They leave their normal job for the time that their role in the shutdown is needed. A maintenance planner might be in this category. |
| **Matrix driven** | People have dual reporting relationships. They report to both the Shutdown Manager and their functional manager. This arrangement is common with shorter shutdowns. Some regular work will be delayed during critical phases of the shutdown. |
| **Function driven** | People report strictly to their Functional manager and handle the project responsibilities as part of their normal work. Such a set-up is frequently part of the organization for any shutdown. A purchasing agent or the buyer might be assigned to the shutdown buying without being relieved of other responsibilities. The agent gets the shutdown work and processes it normally. |
| **Collegial driven** | People are focused on the project but remain in their positions, and project activities center on peer interaction. This arrangement is great for more informal projects. Many companies have such traditions, including the 3M way of developing products, and the old Lockheed Skunk works. The method is not well used in maintenance-oriented shutdowns. |

- There is minimal duplication of functions
- The organization is flexible and can respond to new conditions particularly during the execution phase.
- The size of the shutdown is matched by the size of the staff to manage it. In 1931-1936, only 4 people ran the Hoover Dam project without computers or even cell phones! But there were a hundred people in planning functions from various organizations.
- The organizational structure supports coordination, communication, and quick decision-making
- Roles and relationships are defined and communicated clearly
- Communications systems are sound and complete

**Technology** is playing an increasingly central role in shutdown management and team coordination. At one time, shouting and message runners dominated communications, then we moved to 2-way radios, landline phones/faxes and basic beepers.

Now we add:

- Nextel type radio/cell phones
- Cell phones with camera capability
- Short Messaging
- E-mail and E-mail sent to cell phones
- Intranet Web portals with most shutdown information including updated schedules, Gantt charts, and work lists

- Wireless Networking to laptops and PDAs
- Laptop computer fax software for receiving and sending
- 2-way pagers
- CMMS sending data on jobs and job status to PDAs
- Wireless and Blue tooth PDAs (Palm and Mobile Windows devices)
- Brain implants putting you right on the network in real time (just kidding)

We have all seen that more and better technology does not necessarily mean better communications. There's an intention (you have to want to communicate), actions (log on to the system and download the mail from the server), and responses (you think about what you've read and reply, and bring others into the loop where appropriate), necessary to make a great communicator.

You still have to draft and send the message; you still have to slog through your in-box and you still have to visit the site to know what is going on. But the technologies mentioned ease the road and speed the communications so that managers can be informed of new project conditions in almost real time.

### There are levels of autonomy that are effective at all levels in a shutdown.

Employees take different amounts of initiative based on their experience level and their personal initiative. Also, different managers have dissimilar amounts of tolerance for autonomy. Experienced people are expected to operate at levels two or three (below), particularly when the manager and the contractor or supervisor have worked together before. New people are at level one or two. Certainly the shutdown manager has a three or four relationship with the Policy Team.

The shutdown complexity, difficulty, and danger level itself will dictate the levels of autonomy. A refueling of a nuclear power plant has less autonomy because of the danger, regulatory and legal issues.

**Levels of autonomy**

1. Report facts to supervisor
2. Report recommendations
3. Report intended actions
4. Report action taken

*What level is the right level for your shutdown?*

## Shutdown Team Job Titles

### Manager

**(All shutdowns should have a Shutdown Czar or Shutdown manager).** The manager is given total control of the shutdown. The shutdown manager reports to the maintenance manager or plant manager and to the Policy team (for the shutdown itself). The manager chooses the remainder of the team members, and ensures that the team has all the necessary training and resources (computers, software, communications equipment, etc).

The shutdown manager is the point of contact between the shutdown team and the top management team, which is the Shutdown Policy Team. He or she will refer policy matters to the team in addition to keeping the team informed. The manager is an active part of the Policy team and serves as an expert advisor.

The shutdown manager chairs the meetings that concern the shutdown. Shutdowns are confusing times. The best way to ensure that the confusion is channeled is to have a clear chain of command that is known to everyone involved in the shutdown. The single designated manager (Czar) of a given shutdown is essential in any environment. The person selected might be a manager or an engineer (or both). Working under the Manager until just before execution are the early birds, the planners and buyers.

The shutdown manager is like any good manager that relies on his or her lieutenants in the field to make most of the minute-to-minute decisions. These lieutenants communicate the outcomes, or problems and solutions, to the manager from their actions in the field. The balance between communicating too much detail and too little detail is usually worked out over several events.

The shutdown manager must have wide skills in very disparate areas, but must have major expertise in shutdowns, Project Management techniques, maintenance job planning, basic maintenance engineering, and logistics. A minor in safety, process management, and accounting is useful. Naturally, the shutdown manager also has to be a qualified negotiator, psychiatrist, and confessor.

### Planning manager, Planner or Shutdown assistant

The shutdown manager will require the support of a Maintenance Planner. In a large shutdown, the planning department (headed by the planning manager) is responsible for maintaining the early job lists, job validation, job packages, and the project schedule. Working with the shutdown engineer and plant personnel, the planner will establish activities for all the jobs. The Planning Manager role is a good training ground for a shutdown manager. The planning department turns the work lists into structured projects listing activities, resource requirements, and durations.

The planning manager will work very closely with the shutdown engineer to prepare job packages. The manager also works closely with the shutdown manager and answers questions about the impact of this or that decision on the schedule. They might not have primary responsibility for a safe shutdown (the safety department's might be primary) but the planning team is the first line of defense for working a safe shutdown.

Planning is different from project scheduling. Planning requires significant detailed knowledge of the jobs, techniques, and work conditions. Generally scheduling requires detailed knowledge of project management and less knowledge of specific jobs. But, except in the largest shutdowns, the planners also manage the scheduling functions.

As the shutdowns get smaller and smaller the planner looks more and more like a maintenance planner who happens to be working on a large maintenance job. In a smaller shutdown the job becomes more clerical. The planner completes and manages paperwork and has accounting functions.

The shutdown Planner or Scheduler is (usually) not the same person who is performing day-to-day planning throughout the year, unless the shutdown is a small one, but the skills are the same. The proper method and required procedures, parts, materials, tools, equipment, and access for each job still need to be properly established and assembled in a planned job package. In fact, many of the shutdown work packages can be pulled from the planner library. After the shutdown all the work packages are filed again in the planner library.

## Engineering manager and Engineering support team

These people are the key players for most jobs. Shutdowns require various engineering skills both in the planning phases and in the execution phases. These skills include (if necessary) structural, electrical, chemical, mechanical, and civil engineering disciplines. The group is responsible for all advanced engineering for the shutdown. Much of the danger in shutdown work is the result of inadequate engineering.

All modifications to the plant need to be reviewed by qualified engineers, and these engineers must be given enough lead-time for a complete analysis of the modification. In the UK in 1974, a famous industrial disaster at the Nypro Factory in Flixborough. It killed 28 people and destroyed the plant was the result of a temporary plant modification. That was a long time ago but the same defects are still present in plants worldwide.

### Purchasing agent/buyer

A big part of the shutdown might be buying hundreds or thousands of parts, services, rebuilding activities, contractors, equipment rentals, and other activities that require the skills of the purchasing department. In most shutdowns the bulk of the money is spent for contractors, materials, and rentals.

The Purchasing Agent or the buyer needs to be involved in the early planning stages because so much depends on buying large, expensive, long lead-time, one-of-a-kind parts. Purchasing is essential, but the purchasing representative might service the shutdown without being part of the shutdown team. The representative also might do just the purchasing. Another member of the logistics team might be accountable for expediting parts and materials that are late.

### Shutdown Accountant

To ensure that accounting issues are well managed, large shutdowns will have an accountant as part of the team. These people will account for how all charges are handled. The accountant sets up the accounts for the shutdown and provides instructions for charging of various items.

### Shutdown safety officer

The largest shutdowns will have a person or department tasked with the review of the hazards of various jobs. Such people would be on-hand from the earliest stages to insure safety is built-in rather than added on to the shutdown planning. In a smaller operation the plant safety manager would be lent to the shutdown to assist with the safety aspects of the job. After the plant is shut down the safety manager inspects the project to be sure that safety rules are being followed. Checks would include whether proper PPE (personal protective equipment) is being used, proper techniques are being followed (fall protection) and proper procedures are adhered to EH & S- Environmental Health and Safety department has MSDS-Material Safety Data Sheet for all materials on site.

### Technical support teams

The technical support teams apply their expertise to the shutdown to insure good engineering practices are followed and that the plant is worked on safely.

### Process support teams

These teams assist the planners in making sure the jobs are planned in accordance with the best information from the technical and operations viewpoint. The team's function is to perform engineering, prepare drawings, analyze impacts, give advice, observe work, and report to the project manager, planner, or area owner. The process support team members are responsible for clearances, work permits, quality assurance, and safety inspections. They are also responsible to be sure work can start on schedule by cooling down boilers, cleaning out materials, and decontaminating lines and vessels. The process team is the primary responsible party for the plant shutdown and plant start-up protocols.

### Execution crew

Once the shutdown is underway a new group takes over. Many of the people in this group are the same people as in the planning group who have been shifted over to execution. There are advantages and disadvantages that need to be looked at in a specific plant with the particular personnel. Once the shutdown begins however, the preparation teams take an advisory role and the executioners take over.

### Field Manager

Some shutdowns will have a field manager who is in charge once the shutdown gets under way. The field manager works directly for the shutdown manager. Both these people are in close contact with the customer and directly oversee all activity. Their jobs are in the field, using the MBWA technique (Management By Walking Around). If a contractor is running the shutdown then the field manager is your primary point of contact.

### Shift Boss (superintendent)

This person reports to the shutdown manager or field manager. Forepersons in electrical, mechanical, and other areas report to the shift boss. There is one boss for every shift while the shutdown is going on. This is usually a company person.

### Contractor manager

If the project has a good deal of contract work a team member would be designated as the contractor manager. This manager will interface with the contractors, try to resolve problems quickly, act as lawyer, priest, policeman, psychiatrist, politician, and whatever is needed to get the work done.

### Material and logistics manager

Much of the shutdown consists of managing the logistics of incoming materials, storage of parts and bulk works, prefabrications, and on-site rebuilds. The Logistics team manages all material, all heavy equipment, all hazardous materials and waste, and where everything on the site is going to be kept. The team also manages the site plan showing where everything is stored, and is responsible for keeping everyone up to date with changes to the locations of stored items.

The team also sets up the shutdown storeroom, the storeroom management and control system, and the tool crib. Under this manager the logistics department provides delivery services, police forces (in some cases security would be a function of logistics), shopkeepers, hotel managers, and about 10 other roles. The team stays in close contact with or is part of the purchasing department for materials needed but not ordered.

### Engineering manager

All changes, problems, and engineering decisions are directed to the engineering manager, who should be a team member of the engineering group that put the project together. Again, the greatest safety dangers of a shutdown might be from a minor modification or temporary fix to the plant that "looked " okay. The engineering manager must recognize when outside skills are needed. Mechanical engineers might miss an easily identified hazard if it is civil or chemical in nature.

**Trick to streamline reporting and information**

There are thousands of details in even a medium-sized shutdown, way too many for one person to keep on top of. A person should be assigned to every job on the work list, to and be responsible for all the details associated with that job. The work list is sorted by area and by craft. Additionally it can be ordered by area and by contractor.

| Work Order Number | Short Description | Accountable Person |
| --- | --- | --- |
| 123123 | Replace the major stack blower | J. Jeffries |
| 126566 | Re-line oven #3 | K Carver |

If the accountable party accepts the responsibility, he or she will become the point person on that job, and they will know on a shift-by-shift basis what is the status of their jobs and what are the upcoming problems; in short it is their baby! If 10 people each take on 10 jobs, each having 10 details, that is 1000 details that the manager doesn't have to manage.

# CHAPTER 7

# MASTER SCHEDULE FOR PHASE 1 INITIATION

In each phase (5 in all) we introduce a master list (like the one in this chapter) of things to do, conversations to have, and checks to remember. The goal of all these master lists is to provide input for your own customized master list. Your mission is to have a shutdown list that gets better and more complete over time. Any lessons you have learned should be incorporated so that the list continually improves. The Master list for your shutdowns might only have 30 items to check. But those 30 items could encompass everything for your kind of shutdowns.

Last thought:

*"None of this is rocket science!*
*It is just meticulous attention to detail"* *(Lenahan)*

Much of the work of this phase is redone later in the planning process as details come in and jobs are added and deleted from the work list.

One issue at this stage is that the scope of the work is generally not completely known until, for example, the boiler is opened up or other heavy equipment is disassembled. The scope of the work would include a guess at the amount of deterioration you will find, based possibly on history or experience. The detailed jobs won't be known until the shutdown starts. Obviously the risk increases, if the scope is not known. An evaluation of risk is always part of the presentation to the policy committee. Inspection or NDT (Non Destructive Testing) can be ordered at this time if it would clarify the scope of work.

**Project Broad brush Planning**
**Phase 1 Comment Box**

Discussions at the policy team level. Initial topics include: What problems are we trying to solve? What is the best way to solve them? If a shutdown is best approach then: What is the feasibility, initial scope of work, and budget, approval of approach. At the end of the phase, the feasibility study is complete, the business case study is ready, and documentation for engineering to begin is complete.

This is a time for research, creating a vision for the project and getting approval. The Policy Team creates the mission statement for the shutdown. When the first steps are complete, you will have in hand the beginnings of your Turnaround package. This package will grow in length and detail as each step in the process toward execution is completed.

| Do I need this? | Item | |
|---|---|---|
| | 1. | Talk of a Shutdown gets started and reaches the policy team. Top management appoints a shutdown manager (if there isn't one already designated) to bring some preliminary estimates back to the Policy team. If there is no policy team, one must be chosen by top managers. |
| | 2. | Choose the managers of the shutdown team unless you have a standing team. See that an objective and a mission statement for the turnaround are written and approved by the policy team and distributed all members of the shutdown team and other interested parties. |
| | 3. | If the shutdown team already has an office, display the shutdown objective and mission statements (make them into posters). |
| | 4. | Insure that each member knows his or her role in the overall project. Get training in project management, CMMS (Computerized Maintenance Management Software, if appropriate), and in PMS (Project Management Software) for your shutdown team members as soon as possible. If you are double-teaming to train new people then write a training lesson plan. |
| | 5. | Create the shutdown report. This report will be added to at every step. Some firms start a series of documents on a shared drive on the network specifically for log entries from every person who is accountable for any part of the shutdown to log problems, solutions, activities, opportunities, etc. Web-based BLOGS can be used for this purpose. |
| | 6. | Review unfinished jobs from prior shutdowns and add them into the initial work list (they may be removed later if the priority is not high) |
| | 7. | Gather all data for a scope of work. Determine the preliminary scope of work from the top management mandate, compile an engineering schedule from a review of the total backlog, walk arounds, and previous preventive/predictive maintenance performed. Gather wish lists in formal and informal meetings with operations, process, and other groups. |
| | 8. | Determine general labor requirements, skill sets, and internal versus external labor (contract labor or internal labor) requirements. Determine the elapsed time needed (calendar time) and required time frame. |
| | 9. | Determine the extent of the partial or complete shut down of processes. Determine alternatives (determine if it is possible and desirable to maintain processes by means such as rental generators, air compressors etc.) |

| | | |
|---|---|---|
| | 10. | Determine preliminary protocol or lists on how to shut down the plant and turn it over to the shutdown manager. At the same time, determine preliminary process to return the plant to operations. (Operations is a major contributor) |
| | 11. | Determined whether heavy or specialized equipment will be required, such as cranes, large quantities of scaffolding, compressors, welding machines, or torque wrenches, etc., and if and when they will be available. |
| | 12. | Determine the longest lead-time for supply of spares and other materials needed and the status of materials. The longest lead-time items such as turbine parts will dictate outage timing |
| | 13. | Determine if other plants in your area are scheduling shutdowns during your time frame. This is particularly important in locations with limited contractor or equipment. Some locations have committees of all the major plants & contractors to coordinate shutdown timing. |
| | 14. | Determine what pre-turnaround work will be needed, such as pre-fabrication for the project (particularly long lead time prefabs). In some facilities all work is considered for prefabrication (and you have to prove the case to build it in place). |
| | 15. | Re-read the final report from one or two prior shutdowns and look for opportunities to apply any lessons learned. |
| | 16. | Review the prior safety plan and make changes applicable to this shutdown. |
| | 17. | Set up KPI to measure the performance of the shutdown, to be able to compare it with other plants and with prior shutdowns. Some KPIs can be used by the shutdown team to manage the shutdown while it is going on. |
| | 18. | Estimated Project Cost<br>From the labor estimate, contractor requirements, the material and parts estimate, and the equipment requirements, assign an overall budget for this project. Return to top management with estimates, scenarios, risk analysis, and alternatives, and get the go ahead. Add a factor for emergent work and added work based on history (5% to 40%)<br><br>The total cost exercise includes the cost of having the plant out of service, plant salaries, major risks, and the probability of not coming back up on schedule; in short it is an analysis of the worst-case scenario. There may be a few go arounds, negotiations and conversations to make the cost acceptable and the estimate as complete as possible. |
| | 19. | Create what some people call a level 1 schedule. This schedule has only the start and end dates and other major milestones. |

# Planning Section

CHAPTER 8

# Scope of Work: How to Find and Pick Jobs

When you review your backlog, are there critical maintenance work orders that you can't get to? Are there work orders that require a major process to be shut off or isolated? Are there jobs that are so large that they cannot be handled by normal maintenance planning and scheduling? Are there jobs that require outside contractors because of special skills, special equipment or sheer size? These are all potential jobs for a maintenance-initiated shutdown, but there is a temptation.

There is sometimes a temptation to stuff everything, including the kitchen sink type of job, into the shutdown, especially when the maintenance department is run lean and cannot keep up with routine work, PM work, and emergent demand. Important jobs that could never be completed are added to the work scope. Sometimes this subterfuge is designed to relieve the pressure on the maintenance department and give them a breather, or help the maintenance user by attacking an out-of-control backlog.

Resist the temptation to load on these extra jobs because shutdown work is expensive. Shutdown work is expensive because of the intensive nature of the supervision, excessive use of overtime, and premiums paid for contractors in the special compressed circumstances. Often, it would be less expensive to hire contractors to whittle down the backlog one job at a time, or by the hour, than to add jobs to the shutdown. Adding such jobs might actually cause problems by diverting resources away from the essential shutdown jobs.

A major steel company in the Persian Gulf area started asking the question if any job planned for an outage could be done under normal conditions. After removing these important (but not requiring an outage) jobs, they found their outages significantly shortened (and overall costs reduced).

On the other hand, including some of these jobs might be acceptable if the number of workers needed for the core shutdown varies even after load leveling. Inserting a few such jobs might take employ otherwise idle workers (depending on how the contract is bid. Be careful where added work causes change orders). Another great advantage is inserting jobs that can benefit from scaffolding already erected, large cranes already on site and specialized skills.

The scoping process requires a lot of input from all the plant's stakeholders. In the beginning don't limit the scope or size. Try to collect every detail and winnow the list down as you go. Shutdown job scoping has several aspects:

- Start with any written scope document from top management. This document will be the standard with which new jobs for consideration will be brought and compared (is this new job required or is it even allowed under this scope statement?).
- Solicitation of input via interviews. These interviews are to be conducted with users, operations people, engineers, and other interested parties to obtain their lists of important jobs.
- Scope approval is the stage where top management and stake-holders approve the scope for the turnaround.
- Scope change control is the process whereby new work outside the original scope is added to the project (or rejected).

Once the scope is "managed" (in quotes because it is managed only when the shutdown is complete!) the elements of planning, scheduling, and control can be put into place.

## Sources of shutdown work

- Backlog (CMMS status field = waiting for shutdown)
- From CMMS: Sort and print exception reports for equipment with high costs, high frequency of breakdown, or high numbers of emergency events.
- Routine PM uncovers deferred items. Remember to accelerate PMs (particularly annual and longer term actions) that are due soon, to perform them, if possible before or, if necessary, during, the shutdown. The Preventive Maintenance inspectors are looking at all your equipment on a periodic basis. Their inspection reports (or personal interviews) show areas where subtle deterioration may be taking place. Many

of these items are logical shutdown items.

- Predictive Maintenance reports show where the trend is toward machine decay and destruction. These recommendations could be used to help build the shutdown work list. PdM inspections prior to the shutdown, where a measurement exceeds a known engineering limit such as $+5^{\circ}$ C for electrical circuits, or a velocity of 0.3 IPS (inches per second) for rotating equipment, would indicate problems.
- Special PM and PdM activity for this specific shutdown
- NDT experts can be called in to look at specific areas where there are suspected problems or where there have been problems in the past.
- Equipment history can be in two forms. In some facilities, informal oral traditions are used to pass on information about equipment history. Other plants use formal systems such as CMMS to maintain equipment histories. In both instances, the history might indicate that an item should be rebuilt after a fixed number of hours. Equipment history also shows which components wear out and should be replaced or rebuilt. If you skip the work you face excessive quality problems, capacity issues, and breakdowns, or forced outages before the next shutdown opportunity.
- In manual maintenance departments (no CMMS) the shutdown manager might investigate machines with 'fat folders' (lots of work orders) meaning they have seen a lot of maintenance action and could have unresolved or repetitive problems.
- Organizations with long experience in shutdowns create shutdown checklists. These checklists are developed over periods of years and may include input from similar plants, consultants, and even magazine articles.
- Wish lists, pet projects, suggestions from project improvement teams and TPM teams.
- Interviews and structured meetings with all kinds of stakeholders from workers to top management.
- Engineering projects for improvements of various types.
- Operations projects for ease of operation, quality, improved safety, or other improvement
- Maintenance projects to improve reliability, efficiency, output, or maintainability
- Safety, insurance and risk management inspections.

- Walks through the plant.
- Action Item Lists and campaigns from meetings throughout the year.

The plant's shutdown history files are excellent places to mine for potential shutdown work. In fact, one focus of this book is to encourage the assembly of your shutdown archives in a format that facilitates the search by future shutdown planners. Sources can be both formal and informal and can be from both computerized and manual sources.

### Formalize the walk down tour

Powergen is a Caribbean power generator that operates many small gas turbine generating stations, and outages are scheduled and part of the structure of the business. Before completing the scope of work for a shutdown they have a practice called a walk down audit. This plant tour group includes members of the maintenance department, operations, quality assurance, and others. The group lists everything they see that needs work.

The walk down tours might have checklists for common problems:

- ✓ Safety hazards (potential fall situations, needed attachment points, missing guards)
- ✓ Missing cover plates, plugs, and electrical boxes that are not attached
- ✓ Temporary fixes
- ✓ Housekeeping (complete cleaning in, under, and around machines)
- ✓ Quick checks of oil levels and condition (oil examination, inspection kits or analysis). Look for all kinds of leaks
- ✓ Building maintenance issues (roof leaks), damaged pillars
- ✓ Bad smells or sounds
- ✓ Readings outside or near limits

- ✓ Low oil level (or on the line)
- ✓ Corrosion or damage to structural supports (particularly tankage, silos, reactors outdoors)

## One way to prioritize other work for the shutdown

Mike Brown from NSI recommends assigning 500 priority points out of 1000 to each job (just for making it to the list). Review a list like the one below (adapt it to suit your operation), and add or subtract points to get a final priority (highest jobs win. Keep adding jobs until you run out of time or money). Ask questions such as:

A. Would this job help avoid anything that could endanger the health or safety of employees, the public, or the environment?

B. Will the shutdown work increase output so that the profit will pay for the shutdown in a reasonably short time (especially where we have a demonstrated market for the product)?

C. Is the shutdown necessary to maintain or improve the quality or repeatability of the product?

D. Is the shutdown work required by law, insurance companies, or your own risk managers?

E. Is the equipment to be worked on in the shutdown critical to the operation?

F. Would failure of the equipment in question stop production, distribution of products, or cause a complete shutdown of the facility?

G. Will the problem get worse by the next shutdown if it is not done?

H. Will the job greatly increase in price if it is delayed?

I. Is the capital investment high in the equipment to be worked on (we might want to protect our investments)?

J. If the job is not done, is plant capacity compromised?

K. Can the load be easily shifted to other units, plants, or work groups?

L. Does the normal life expectancy of the equipment without the shutdown exceed the operating needs? If this is true, the shutdown may be a waste of money.

M. Is the cost of shutdown work greater than the costs of breakdown and downtime (shutdown might be a waste of money)?

All these lists are merged into a master list and converted to action oriented work orders. The work orders are subject to validation before making the final cut to the shutdown.

## Strategic shift due to global competition

Benard Ngesu, a maintenance leader from an Eveready battery plant in Kenya reports that his plant used to fix a date for the shutdown and when it came they would rebuild everything, whether needed or not. He goes on to say "you can imagine the chaos during start-up. Usually it would take days before equipment could work well again –obviously due to infant mortality (start-up problems) ... That was a long time ago."

Now things are different "with increased competition we began to think about ways of cost control and maintenance was identified as an opportunity area. Shutdown time, and problems with start-up of the equipment were points of focus. Most of the work that used to be done during shutdown has today been taken over by scheduled maintenance (corrective jobs resulting from PM and PdM inspections that is planned and scheduled), so that only jobs that cannot be done during normal production need to have a shutdown. Planning is a major consideration in all shutdowns I'm involved in today."

## Position yourself for the future

Once a plant is shut down and safe you have a golden opportunity to do inspections, take measurements; make drawings that would provide useful information for subsequent maintenance, and shutdown work.

Everything from gathering hard-to-get-to name plate data (one trick is to lay a piece of thin paper over the name plate and lightly go over it with a pencil to create a rubbing of the plate), counting gear teeth (useful for vibration analysis), to NDT tests on tubes or piping could be useful. Sampling that is normally hazardous, sketches of wiring, or identifying old spares, could give you an edge on the next repair on the unit or shutdown.

Mr. Mok Kheng Heng described one interesting strategy practiced at Schering-Plough in Singapore. During their shutdowns, the OEM (who actually built the equipment) come onsite to repair and inspect equipment. While they are making repairs with the equipment opened they decide what repairs will be needed for the next shutdown. The OEM vendor creates proposed work orders and complete parts lists with some level of justification. The vendor also makes any measurements needed to do the work. This work forms the beginnings of very accurate job plans for the next shutdown, which is a tremendous advantage for the future.

### Road trip

The process of the shutdown is just like a road trip from your home in Chicago. To start the process the executives of the family decide to visit New Orleans for Mardi gras. Phase 1 is the initiation of the project. The policy team sets the goal and assigns a budget. All the stakeholders argue for their pet (important and necessary) projects within the structure set out by the policy team. For example Jim wants to stop in St. Louis for Ribs (almost on the way to New Orleans and will pass at mealtime- approved). Sue wants to visit her friend in Knoxville (Off course-disapproved).

Phase 2 is the planning process. The planning team jumps into action, fires up the computer and logs on to Mapquest.com to get an estimate of the distance and route. The team might even predict where the family will be at what time (create a milestone), pick restaurants, and make reservations for sleeping.

Phase three is the drive itself. Once you start driving you might monitor your progress every few hours against your plan. Driving is a different skill set from planning the trip. As a driver you act and react to detailed circumstances, debris in the road, crazy drivers, police, etc. As a driver you are responsible for delivering you and your family safely to your destination. A crash by the driver will spoil the best plan!

A good plan makes the trip easier. If the plan was based on a map of Oregon, the best driving in the world won't help. Good driving makes the trip safer, quicker, and cheaper.

Phase 4 is arrival and unpacking your car in warm New Orleans. Phase 5 is the stories you tell your friends when you return and for years to come.

CHAPTER 9

# WORK VALIDATION AND PREPARATION FOR PLANNING

The first step of planning after the work is identified is to purify the work list. The word purify is used because once the work is approved it is necessary to remove duplicate, erroneous, and misstated jobs, leaving only the pure work that is totally cleaned of work order debris. CMMS (computerized maintenance management systems) are notorious for accumulating debris. A regular program of backlog purification (validation) is a great idea and is to be encouraged. The work must be whittled down to essential items only. This process is called validation.

After the scope document is prepared with the list of all approved work orders, each work order has to be validated. Validation is an investigation of all jobs on the shutdown list. (Partially adapted from *Turnaround Management* in resource section)

Does the work have to be done during this shutdown (or any shutdown)? Work done during shutdown is more expensive due to the cost of the premium time and extra control needed. There are exceptions but the question needs to be answered.

If the work order is from the backlog on the CMMS, does the work still have to be done?

Is the work a duplicated job or is it a small job that is part of a larger job already on the schedule. Or worse yet, is this a job that will be undone by another job later in the shutdown.

Simplify the individual work orders where possible. The work content should be entirely clear so that the way the work order is written is important. Certain simple wording will add work which might be unnecessary such as "replace valve" versus

"Examine and replace valve if needed"

Beware of normal routine and corrective work that has been transferred to the shutdown list

Avoid nice-to-do work for which there is no business justification.

If you have spare resources a few days after resource leveling you might take this work on but be careful what you promise.

Is there any work being done as a custom every shutdown without adequate analysis? Traditions are great but in shutdowns there had better be a better reason than "we've always done it that way." In some instances, an expert might have to be brought in if no one in the plant truly understands why something was done in the first place.

Be alert for desperation or blackmail work (that is, work added to the list just before start-up) on an "oh by the way without doing X the process won't work." This extra work must be accounted for and highlighted for the policy team.

**Validation Process:**

- Check for duplication, approval, and that the work requested is written accurately
- Challenge the need to perform the task, either at all or during this shutdown
- Analyze safety, quality, engineering, material, and resource requirements
- Look for opportunities to make economies such as shared set ups (scaffolding, plastic sheeting)

**Validate: agree on final wording and add to work list.**

Running a class in how to write up the "work requested" field on the work order is strongly recommended. Many people and companies have trouble with this chore (just pick up a stack of work orders and read them as an outsider would). The issue of how work requested is written up is important for day-to-day maintenance jobs and doubly important for shutdowns. Frequently the requestor knows what the mechanics have done in the past and writes up the request that way (such as replace valve 5J since valve 5J was always replaced in the past).

But this time the work needed is different, or we want to get to the root cause of why 5J seems to be always leaking. In other areas we are doing a bigger job like replacing all the valves on line 5 with a more robust technology (and 5J would be replaced in that work order). Work validation ensures that the job is written up in the most useful way and also looks carefully for jobs that include other jobs.

## Packaging

Once validated, the work has to be analyzed so that it can be made into bid packages and into categories so that adequate planning can be done.

Work to be contracted is packaged by trade, skill, or machine, depending on what type of contractor will be used. For example, the jobs relating to the boiler might be packaged for a boiler contractor. All electrical work might be packaged for an electrical contractor.

Categorization is used to divide the work into major categories (Lenahan): major tasks, minor tasks, bulk works and construction.

**Major Tasks:** These tasks are hazardous, complex, unknown, high work content, and large numbers of skills, multilevel work, or of long duration. The planning techniques would include: Deep WBS (work breakdown structure), step-by-step breakdown, resource loading by shift, schedules for heavy equipment and large tool use, special drawings. The plan might include detailed scope of work, lockout tag-out with every isolation needed, procedure for welding, connections, testing and NDT, and instructions for return to service. Extra effort is made to collect all relevant documents including drawings, wiring diagrams, manuals, SOPs, safety plans, any special clean-up instructions, etc. Finally the plan would include a closure document in which the craftspeople could add any resources used but not on the plan, plan changes, and other feedback to the planner.

**Minor Tasks:** These tasks are smaller and simpler than the major tasks. The hazard level, complexity, and work content are all moderate or low. Minor tasks are regular maintenance jobs that happen to require shutdown, and are familiar to the planning team and to the work crews. These tasks do not require (much) heavy equipment or specialized skill sets. A work order with labor requirements and parts list would be expected. These tasks do not usually need to be broken down further into sub activities, and a minimum of extra services will be needed.

**Bulk works:** These are large groups of similar or identical, relatively simple jobs including work on valves, small pumps, steam traps, burst disks, etc. These jobs are scheduled as blocks of work. One typical job might be worked up for reference. A large group of people might be involved at one time. The difficulty is to ensure that each item is in fact done.

**Construction work:** Standard construction jobs include buildings, tanks, pipelines, and retention ponds. These items are generally designed by engineers and architects and are bid conventionally with fixed price contracts.

## How to Determine Planning Lead Time

1. Estimate the amount of labor hours for the whole shutdown including contractor hours.

2. Estimate the planning time to be a range of 2 to 10% of total hours. The exact number of labor hours depends on:
   a. The complexity of the jobs – higher complexity = higher percentage
   b. Expertise of the crew in the specific type of job (more expertise = lower percentage). A contractor that does this work everyday can supply some or substantially all of the planning.
   c. Overall size of the shutdown (small ones need higher percentages to cover the planning overhead such as setting up files, making up lists, preparing preliminary budgets, etc).
   d. Presence of extreme hazards (more hazardous = more planning)

3. Figure on needing 1 person-year for every 1700 hours of planning time, for each dedicated full time planner. Expand the planning lead-time if you can't get full dedication.

**Example**

A shutdown in a chemical plant might entail:

- ✓ Replacing several thousand feet of piping, fitting of different sizes with insulation
- ✓ Removing and replacing several reactors
- ✓ Removing a large heat exchanger
- ✓ Repairing 75 boiler tubes

Let's say that the total hours are 22,000. The job is mostly repeats of other jobs done throughout the years. The hazard level is moderately high and the complexity is medium to complex (it includes three-dimensional work sites).

The 2 to 10% would yield between 440 and 2200 hours for planning this shutdown. Given the circumstances a planner might settle on 5% for 1100 hours or 7-8 months of full time planning. If the lead-time on the scarcest part is greater than 7 months then the planner might break up the planning to suit the lead-time.

## One technique to add all the individual Job plans together

### Work Breakdown Structure (WBS)

The WBS has its roots in the beginning of scientific management at the turn of the last century. Fredrick W. Taylor (considered to be the father of scientific management) designed the precursor to the currently understood WBS.

"The Work Breakdown Structure is a hierarchical list or diagram representing all the tasks that must be completed in order to complete the project." (Project Management Mentors - see resource section).

The WBS is already designed for fields such as architecture and construction, and for many government programs for NASA, and the military. One definition that is important is the depth of the WBS. It could be 2, 3, 4, or more levels. At the end, at the lowest level is the terminal element. A **terminal element** is the lowest element (activity or deliverable) in a work breakdown structure; it is not further subdivided.

### WBS Technique for a shutdown with discrete jobs

As we have discussed before, unlike most projects, the shutdown is made up of related, quasi-related, and unrelated jobs. This condition complicates the assignment of work breakdown categories. It is recommended that the planner:

- Print out all the jobs onto sticky notes (or use adhesive tape).
- On a white board, wall, or large sheet of paper, organize the sticky notes into an ad hoc structure that makes sense for that shutdown
- Write out the structure and assign the people to be responsible for each major line.

WBS elements are usually numbered, and the numbering system may be arranged in any convenient way. If a WBS is extensive, and if the category content is not obvious to the project team members, it may be useful to write a WBS dictionary. The WBS dictionary describes what is in each WBS element, and it may also say what is not in an element, if that is unclear. Here is a sample of a WBS dictionary description:

WBS Element 1.5.4.5. - Systems Integration Test Equipment Planning - This element includes the effort to identify requirements and specify types and quantities of test equipment needed to support the System Integration and Test process. It does not include the design or procurement of such equipment, which is covered in Element 1.5.4.6. From *James Chapman, PMP*

The WBS facilitates estimation by breaking the work into manageable chunks. The tasks should be broken down until they can be well understood, it is easy to tell if they have been completed, and they can be assigned to one person or crew.

## Use the WBS to assign responsibility

Think about the major areas of the shutdown based on the Work Breakdown Structure and assign responsibility from that structure. The work breakdown structure is discussed in more detail in a later section. Point people for level one; level two (or whatever is appropriate), can be assigned.

| | | |
|---|---|---|
| 3.3 | Shutdown Engineering | Asfar Kingston |
| 2.31 | Civil engineering for site | John Claxton |
| 2.32 | Structural engineering | Fran Mullins |
| 2.321 | Calculate supports needed for Reactor 6-H | |
| 2.322 | Review scaffolding shop drawings | |
| 2.323 | Engineer bridge for section 12 | |
| 2.33 | Electrical engineering | Kiki Shana |
| 2.34 | Process engineering | Bart Dalton |

Engineering (all of 2.3 rolled up) is the responsibility of Mr. Kingston. He might be the manager, where all the other engineers are direct reports or he might be the point person whereby all activity statuses are managed. This assignment of responsibility is important because the party might be responsible for knowing what's going on, for getting additional resources, and for resolving conflicts. Depending on the structure and agreements, the responsible party might also control or supervise the job.

All the civil projects under 2.31 would be the responsibility of Mr. Claxton, and the structural projects (2.32) would come under Ms. Mullins. Ms. Mullins would be the point person for all the activities under her section in the WBS. Again she might or might not supervise them.

The whole shutdown, including trade work, could be divided in this way. In some areas, the responsible party might do the task breakdown, the job estimates, and scope of work for their specialty.

CHAPTER 10

# Individual Job Planning

**Plan:** (Defined by Webster as a method for achieving an end). The end might be a pump that transfers acid into the process or a boiler that produces the requisite amount and pressure of steam. In maintenance planning, the planned job packages are detailed methods to achieve the end (that is the repair or replacement). Notice in the definition the word method. Planned work has the method to some degree worked out in advance.

## The Planned Job Packages

The Planned Job Package is to be prepared for every independent job (for more details about the Planned Job package consult the book *Maintenance Planning, Scheduling and Coordination* listed in the resources section).

As mentioned many of these planned job packages are already filed in the planner's library. Job plans from all routine maintenance jobs, as well as every shutdown job done in that facility, should be in the planner's library.

*Idea for action:* Be sure that after additional planning is done, the new, improved, or more detailed plans, make their way back to the planner's library for reuse. The planner should also spend some time cleaning up these plans before filing them.

## Idea to improve Planning in a multi-plant environment

The shutdown planners as well as the routine work maintenance planners should be in loose association with each other. Although these planners work for their own organization they are also members of a planner's group. In the best organizations, this group will meet for training and discussions to improve the state of the planning art in their organization. The important thing to make everyone's job easier is the sharing of job plans. Of course with large scale CMMS; it is easy to share the plans (such as they are). Having a starting point from a job plan for a similar type of asset could save hours.

*Keep generic and unique plans.* A generic plan might be to remove and replace any motor from a pump. A unique plan would be an adaptation of the generic plan that adds in all the specifics for a particular pump. It might include special procedures because of the type of pump, special tools, special safety instructions, and specific part numbers. Once a good generic plan is available, the planning function becomes fill-in-the-blank.

### Engineering

You have to decide if engineering is required by the scope of the job. If engineering is required then the planning should be delayed until the engineering has been completed. Because of the possibility of introducing an unintentional hazardous condition, engineering should look over most, if not all plant modifications and especially temporary situations. Replacing like with like might not need engineering review. Some of worst accidents in modern times were the result of seemingly innocuous plant modifications going bad.

In oil refineries, engineering is extremely important to both safety and integrity of the process. The BP T&T refinery actively manages all drawings going out for bid, shop drawings coming in from contractors, and modifications to the plant. Each version is audited, tracked, and managed. The firm can reconstruct the steps taken for engineering all shutdown work.

## Steps of the individual job Planning Process

**Developing a job plan requires a logical, step-by-step process.**

After defining and describing the job objectives and scope, list the steps to be performed; thus defining "what" is to be done. A quick look at the planner's library will determine if the job or a substantially similar job has been done before, and if you already have a job plan on file. A search of the planner's library will also determine if a generic plan or a plan from another plant is available.

If the job is not familiar to the planner, a visit to the job site and a discussion with the requestor might be the next step. Even if the job is familiar to the planner, sometimes a visit will clarify job details that might make a difference during the shutdown. For example, there might be additional obstruction since the last time the planner visited that asset. Pinning down exactly what is to be done is sometimes half the battle.

How important is this job in the overall scheme of things? Is it an essential job that must be done, or a nice to get done job? Assign a priority to the job to relate it to other shutdown jobs (which might be different from its priority during a routine maintenance week).

Consider the advice of Mike Brown "visualize the job and how it will proceed. Visualization means forming a mental picture of the job and the people working on it." He continues with advice that it is important to get the right sized job steps. Research on labor estimates for a wide variety of jobs compared with actual records shows that the best length for an individual step is no less than 15 minutes or more then 4 hours.

His other excellent advice is that many jobs are symmetric in that the steps to remove the item can be reversed to install the item. Examples include removal and replacement of motors, pumps, fans, compressors, etc. Recognizing this symmetry could save some planner time.

Taking the physical location and spaces around the equipment into consideration, plan the manner in which the work is to be accomplished, thus defining "how" the job will be performed. Some of the how is answered while breaking the job into steps. Include ideas about how many people can work efficiently, and consider the movement of materials into and out of the area. Determine the minimum and maximum crew sizes as well as the 'best' or optimum crew size. Look above and below the level of the work and note any hazards that might extend the estimate for duration of the job.

### Known Hazards that impact job time

Some hazards to be aware of include: working at heights (over 10 feet), working near moving machinery, working near energized machinery, working above or below other jobs, work requiring full PPE (personal protective equipment), excessive heat or cold, and bad weather.

Establish duration and manpower needs required to *"perform"* the work. Determine the skill sets needed, and any special licensing requirements (i.e., certified welder). Clarify the sequence and timing of skills required throughout the job.

Consider whether your firm has the skills to do this job? Will the work be done in-house, will a contractor be called in, or will some combination be used to do the job? These steps define "who" is to do the "what." Deciding on contracting might be delayed until all or most of the jobs are planned and you have a better picture of the size and scope of the shutdown.

Next, identify all spare parts, materials, consumables, special tools, PPE and equipment necessary to do the job. Determine if the items are standard stock items. If not, determine where and when they can be obtained. The plan must include the conclusions from this research.

Step back from the immediate site. Will there be any difficulties in delivering the materials to the work site? Can you fit in the crane; can a worker maneuver the part? Consider the issues presented from when the spare is off loaded from the truck or rail car to when it is placed to bolt up. Predetermine trouble spots and work through solutions.

Summarize the required budget for the job, thus defining "how much" it will cost. The most accurate way to determine a shutdown budget is to start from budgets of each of the jobs. We recommend that all jobs be estimated and be given a budget, even for routine maintenance (most CMMS will facilitate this step).

The job cost would be based on the labor hours times the labor rate from a typical contractor. Job steps, skills, crew sizes, and durations have already been determined. All prefabs, parts, materials, and supplies should also be priced and added to the estimate with material overheads. The need for specialized equipment can be put aside unless this is the only job that requires the equipment. If the prior steps are complete this step is simply arithmetic.

Then, determine accessibility of essential reference manuals, drawings, diagrams, and other materials, and include them or references to them in the Planned Job Package. Some CMMS allow links to documents on work orders. At this writing the linkage between CMMS and document management systems is still primitive. Most organizations also do not have these documents in computer accessible formats. We are certainly moving toward this merger of the two types of systems. To aid this process many OEM are providing O & M manuals on CD or providing for downloading from the Web.

Consider the hazard introduced by the job. Is a JSA (Job Safety Analysis) in order? Will any clearances be needed? Is there going to be flame cutting, welding, confined entry, or any other activity that will activate the safety permit system?

The process of job planning encompasses verification of all aspects of the job to be done as well as identification of the various input resources (material, manpower, and equipment) required to complete each job.

Recognition that jobs often go bad is part of good planning. Risk assessment and contingency design is part of the planning process. Can the risk be eliminated? If not, how can its impact on the shutdown and organization be minimized?

The first part of the process is job assessment and scoping, which should roughly follow the outline below:

## Job Assessment and Scoping Checklist

*(From Maintenance Planning, Scheduling and Coordination in resource section)*

| |
|---|
| • Confer with the requestor to clarify the desired result. By clarifying the end objective, the means to get there can often be simplified or the job can be expanded to solve several seemingly unrelated problems at once. |
| • Refine the description accordingly (the wording of the work order is important) |
| • Clarify the specific job location (building, floor, bay, etc.) |
| • List what needs to be done (job content) |
| • Define start and finish points (job scope) |
| • Finalize priority |
| • Visualize job execution and outline the steps and requirements |
| • Record the steps necessary to execute the job |
| • Prepare sketches or take photos to clarify the intent of the work order for the assigned mechanics or simply as a reference during detailed planning |
| • Take minor measurements (exactly). Complex measurements (i.e., for fabrication) should be left to the assigned technicians. |
| • Determine required conditions. Must this job be coordinated with Production or another group? |
| • Is it necessary for equipment to be down (major or minor?)? |
| • Define involved successor and predecessor processes (what is impacted by this job and what this job impacts) |
| • Determine if other equipment or adjacent areas will be impacted by performance of this job? |
| • Check for safety hazards, perform a JSA or have one done if needed (this can also be completed as the job steps are thought through). |
| • In some plants operations hang tags where requested work is located. The tag number is written on request. |

## Job Research

An important step is to search through labor libraries, reference files, and the Maintenance Technical Library to determine if the whole job or parts of the job have been planned previously. You can also use the files to fill in information, knowledge, and reference gaps. Always use the information available to avoid redundant effort. As mentioned, check sister plants for similar or generic job plans.

Consult the procedures file to identify necessary tag outs, safety inspections, fall protection anchor points, fire watches, and standby positions associated with open flames, ladder and scaffold use, vessel entry, etc. Safety must always be a primary concern within the planning of shutdown work. The best safety strategy is to eliminate the hazard; the second best is to build safety into the job plan. Planning is critical to safety and loss prevention, because the planning stage is the logical time to think about and incorporate safety matters into the Planned Job Packages, before the pressure to finish the job comes into play

## Detailed Job Planning

- Dig through the library to find a similar job to establish a starting point for your planning activity (if possible).
- Visualize people doing the job. Try to see how the work would proceed.
- Select and describe the best method to accomplish the job (consult with experts if necessary). This 'best method' can be an area of contention so if the job is an important element of the shutdown, a round table discussion with experts (including some people who have done that job or similar jobs) before starting planning is desirable.
- Consider lock-out, confined space entry, fall protection and all other safety factors
- Determine the job sequence by specific and logical tasks or steps (again the most accurate job step is between 15 minutes and 4 hours)
- On complex jobs, use PM software to determine the optimum schedule and coordination of crafts and crews.
- Determine labor resource requirements including
  - Required crafts
  - Detailed skill sets if craft is not detailed enough (some crafts have sub-sets such as instrumentation as a sub-set of

electrician or high pressure vessel welding as a sub-set of welder)

  - List license requirements (code welder, asbestos removal)
  - Minimum, optimum, and maximum crew size (you may want or need to crash the schedule or run the job with a smaller crew. It would be nice if this aspect is already thought through)
  - Sequence and timing of different crafts (you don't want people standing around waiting for their piece of the job to be ready unless it makes sense.)
  - Establish labor-hours for each task of the job sequence
  - Determine whether special or extra allowances are required for the specific job being planned. Some reasons for extra allowances might be working at height, heat, cold, near electric, moving machinery, etc.
  - Determine whether contract resources are needed can be done at this stage or later, when all the jobs are aggregated.
  - Determine time is needed from operations for cleaning, dewatering, or decontamination (this type of work should be done as part of the plant shutdown but may not need to be done for a specific work area)
  - Determine time needed for cooling down and heating up, to make the asset accessible for work.

- Prepare the Bill of Materials for the job, listing necessary materials, and parts required. Be as specific as possible with part numbers and quantities. Establish an acquisition plan. Determine what items are in stock, and whether to make the part or buy it? For direct-order items (not authorized for stocking), the planner prepares the associated Purchase Order Requests or material requisitions.
- Determine what in-house or outsourced pre-fabrication, is needed. Create Work Orders (or purchase orders) for fabrication. Get estimates for lead-time or look at actual lead times for past jobs that were similar. Consult with the prefabricator if indicated.
- Determine any rebuilding requirements. Create Work Orders for rebuilds. Source the rebuilding. Prepare paperwork for the rebuilding if it will be outsourced and moved off-site.
- Identify and quantify bulk works (for example mass rebuilding of valves)
- Is large equipment needed? If rentals are needed, prepare

Purchase Order Requests with Work Order references for equipment rentals.

- Identify special tools and equipment required, including safety items. Identify where large numbers of normal tools will be needed (like 100s of torque wrenches)
- Consider how to get parts, people and equipment to the job location, together with ladders, scaffolding, rigging, cranes, and other heavy equipment. Consider if there will be storage problems when the items are not in use. Is the ground safe for the loads imposed by heavy lifting equipment?
- Coordinate related work of other groups if important by preparing Cross Work Orders, or add unique Tasks on the same WO (Work Order) if only minor support is needed
- Consider disposal issues (for both liquid and solids, asbestos, intermediate products, oils, and other contaminants)
- Estimate total cost in terms of labor, materials, and external charges
- Perform a risk analysis for what can go wrong. Could the job scope suddenly change, could the job uncover significant additional work? If so create a contingency plan.
- Conduct a risk assessment or JSA. What is there about this job that presents a risk to the whole shutdown, to the personnel, the public, or the environment? Have a contingency section in the plan where you outline the risk and if it cannot be eliminated how you plan to manage it.

## Ways of Estimating Maintenance Jobs

There are several ways to estimate the duration of maintenance jobs. The importance of good estimating cannot be overestimated (but sometimes it is underestimated!). The object of estimation is to predict the duration and labor content of each job within plus or minus 15% of the actual cost.

Project management techniques can handle some inaccuracy in the estimates. However, good estimates make good plans, good plans make good schedules, and good schedules make good shutdowns.

Be sure you distinguish the difference between the numbers of labor hours to complete a task verses the length of time the task will take. The length of time (duration) times the crew size equals the labor hours. This is a very common mistake when discussing jobs because the planner is usually dealing with total labor hours and the project team is more concerned with durations.

### As you can imagine, scope is essential to the estimate

Estimates depend on a clean, accurate scope of work. The better defined the task the better and more complete the estimate is likely to be. Estimates can go wrong when the scope of work changes dramatically. If the initial job scope to change a simple bearing morphs into replacing all the bearings plus the shaft, and impeller, the scope change dictates that a new estimate should be made. Judgment must be used.

### Where do estimates come from?

The first place to look for estimating data is your own history with similar jobs and in shutdowns done in the past. These data are described below as Historical Standards. Your skilled workers might remember the duration and crew makeup of similar jobs. Their estimates might be surprisingly accurate.

Some industries have worked together to create estimates for common maintenance work. A consortium of offshore oil drilling companies has developed a set of estimates for jobs on oil drilling platforms. Both the contractors and the owners use the same estimating basis. The standards of the consortium are applied on oil drill rigs worldwide. In different parts of the world the contractors use different factors. These factors include allowances for local skill levels, local motivation levels, difficulty of working in the location (200 miles from shore is different from working 2 miles from shore). In the gulf of Thailand, a factor of 3X-4X is used. Even in the Gulf of Mexico the factor is 1.3X.

There are several places, both internal and external, where you can get estimates for shutdown jobs. Each of these sources has advantages and disadvantages. Some of the most common sources are listed below.

### Guess, SWAG

Although they may be educated guesses, most installations begin with guesses. A guess is based on nothing. An estimate (one step up from here) is based on some judgment or other information. Guesses are the least costly, least time consuming, and least accurate.

The disadvantage of guesses is that they are personal. They reflect the personal opinion and experience of a particular supervisor, planner, or technician. Different people guess differently. Guesses are tough to distinguish from quick estimates. One person might be basing the guess on a lifetime of knowledge (making it more of an estimate) and another might base it on nothing.

If the planning system has feedback (where the planner gets to know how long the job actually took compared with their guess) then the guesses could improve pretty quickly and become estimates.

## Estimates

Skilled planners and trades people can look at a job, mentally compare it with jobs accomplished in the past, and come up with an estimate. We make estimates all the time. Estimates are different from guesses in that they are based on experience. They suffer from the same problems as guesses in that they vary in accuracy with the experience and skills of the estimator.

Great mechanics often make terrible estimators. The skill to do work does not always translate to the skill to estimate the time needed to do the job. There are several aids to estimating that improve accuracy including slotting, and analytical estimating tables.

Some estimates are based on some quantity, as when a steel fabricator estimates based on the number of pounds of steel or an excavator estimates based on cubic yards of excavation (also called scaling).

## Analytical Estimates

Analytical estimates look at the pure work content for each job and add factors for travel, fatigue, and working conditions. There are tables for travel (how long it takes to go 100 feet, 500 feet, etc). The number of trips to obtain parts is allotted, based on the length of the job. Factors are added in for fatigue (after so many hours), and for working at height (add 15% for 12 feet).

Analytical estimation is moderately time consuming but results in excellent estimates. If an organization has an effective analytical estimation program in force for a year or more then their labor libraries will be full of useful information when it comes time for a shutdown. Consult *Maintenance Planning, Scheduling, and Coordination* in the resource section for details on how to use analytical estimating.

### Job Slotting

Job slotting is an effective work measurement technique because it is easier and more accurate to compare jobs than it is to estimate them from scratch. Slotting uses well known jobs of different durations as standards or slots. Each of the standard jobs has a well thought through scope, and planners and technicians have agreed upon the duration.

The job you want to estimate is compared with the standard jobs in the slotting table. It is moved up the slotting table until the planner determines that it is shorter than one slot and longer than the slot below it. The estimated duration standard for the target job is set as half way between the two slots.

| Job | Description | Slot |
|---|---|---|
| Changing a 110V outlet or switch | No problems, wires adequate length | 15 minutes |
| Removing and replacing a fractional horsepower motor | Just bolt up, wire, and go. No alignment issues | 1.0 hours |
| Install 220V branch circuit 100 feet | Just hang | 4.0 hours |
| Replace breaker panel | 28 circuits, standard mounting, no problems | 8.0 hours |

### Historical Averages

The labor-hours charged against previous work or individual jobs are recorded and accumulated. They are then averaged after elimination of skewed highs and lows. The resultant averages reflect size and condition of the facility, condition of the equipment, skill level of the maintenance work force, and current state of job preparation and materials support. Because the work order system is the source of historical averages, it is often difficult to obtain reliable data on which to calculate the averages. In too many CMMS implementations, work content is charged to whichever work order is handy, not necessarily the right one. Another disadvantage is that the standards include all the lost time typical of that plant.

### Universal Maintenance Standards

Predetermined Motion Times, Time Study, and Standard Data evolved into Universal Maintenance Standards (UMS) and even these have faded from common usage. Although they are the most accurate method, such standards are too time-consuming and expensive to set up as well as maintain. Each job is studied by an industrial engineer and analyzed for best practices. Only consider this technique if you have to change thousands of filters, rebuild tens of thousands of valves, or perform some production like job. Because of the problems, these standards are not generally recommended.

### Construction Trade Estimates

Construction Trade Estimates are developed for contractors to use when bidding on construction jobs. While they are not recommended for in-house use for maintenance they can be used as guides. There are some specific estimating sources for the chemical industry that are better than others.

The reason for the lack of standard methods is that construction is not among our most efficient industries and the estimates reflect that level of efficiency. The standards also include engineering safety factors in the interests of the bidder, and most relate to construction jobs rather than maintenance jobs. If this is the only guide you have access to, use it by all means. One idea is to compare the estimated time for a few known smaller jobs to the actual time taken for those jobs. Use the ratio between the two on unknown jobs to make the published estimate more relevant to your conditions.

### Flat rate manuals

Some industries (automotive, mobile equipment, HVAC) have repair times, techniques and steps that are well documented. When you buy a truck, for an extra $100 or so you can get a manual of repair steps, times and other useful information. Consult the manufacturers of the equipment or components for information on trucks, construction equipment, seals, bearings, etc.

These times tend to be very well thought through but they assume a level of tooling and work conditions that you might not be able to duplicate. The OEMs use the flat rate manuals to reimburse dealers for warrantee repairs. You can be sure the times are quick (maybe too quick if you have older, dirty, or rusty equipment). Some dealers pay their mechanics based on these rates times their pay rate for all the jobs they complete that week.

### Engineered Performance Standards

These standards are a special case of a flat rate manual that was compiled by the US Navy in the 1970s. They employed an army of industrial engineers to time study all maintenance work performed on Navy bases. The result was a library of manuals with accurate estimates for work such as painting, carpentry, and electrical work, even standards for pier building! These Standards are available from the Superintendent of Documents.

## Planned Job Package

During Job Preparation, the Planner assembles and documents all the above planning efforts within a "Planned Job Package." Any factors that may delay or hinder effective job completion should be anticipated and steps taken for their avoidance. Different jobs would involve different levels of complexity and completeness of their planned job packages. A minor job might only need a work order with a bill of material and appropriate clearances. A major job might require 50 work orders and all the elements shown below. The Planned Job Package could include:

- Planned job package table of contents (for larger jobs), or cover sheet
- Detailed Work Order (or work orders if the job is large) spelling out the work to be done
- Job Planning Sheet with Sequenced Tasks detailed by craft and skill level. Contractor as well as in-house resources must be included.
- Lockout and tag-out, confined space entry, instructions if they are not included on the work order.
- JSA describing hazards, Safety requirements and PPE requirements.
- Requirements for decontamination, dewatering, cooling, and removal of product.
- Details of all pre-shutdown fabrication. Estimates of the lead-time for the prefab work. The plan should be to perform as much pre-shutdown fabrication and other preparation as possible
- Duration and labor-hour estimates for each task. When beneficial, (particularly for longer, more complex, or critical jobs) a Gantt or CPM chart should be prepared to convey trade sequencing and simultaneous tasks to the assigned crew(s).
- A Bill of Materials including availability, commitment and staging location. The listing should distinguish between authorized stock items, direct purchases, indirect purchase (contractor will buy), in-house fabrication, and outside fabrication.
- A list of all specialized tools, machines and heavy equipment.

This list should include everything from cranes, to line freezing rigs, to scaffolding. It also includes power tools that are common but not normally carried by the trades such as porta-band saws, threading machines or impact hammers. The list also includes standard tools that are needed in large quantities.

- All clearances and all required permits (both governmental and company) completed to the point of safe feasibility. These documents might include flame permits, confined space entry, open line permits, valve close/open permits, discharge permits, building permits, etc. Of course, the final lock outs must be made by the responsible mechanic and equipment operator
- Outside specifications (Standards) that apply to this job such as ANSI, ASME, etc.
- Other reference documents that the assigned crew is likely to need, such as prints, sketches, photos, specifications, sizes, and tolerances.
- O & M and other manuals or sections of manuals as required.
- A Site Set-Down Plan (for major tear downs).
- Sign off sheets for job completion from contractor, data entry, safety, and others as appropriate.
- List or hold points for quality or safely inspection.

# CHAPTER 11
# RISK MANAGEMENT

Shutdowns are big events in the life of a plant. The size itself without any other factors constitutes a risk for the organization. Major risks are identified for the Policy Team at the outset of Phase 1 when the shutdown is first being discussed.

The challenge is to balance the tendencies to go off the deep end on the one hand and pretend there are no risks on the other. Being overly pessimistic tends to paralyze the team. Being overly optimistic blinds the team to risks that can easily be mitigated or eliminated.

Shutdowns may have elements that have never been done before by your organization (or perhaps any organization) that create risk for the organization. By definition, these large projects include jobs that are unique (even if they were done before they always seem to be different). Because these projects are unique they carry a higher risk than repetitive work.

What is the risk if a power plant is delayed a few days or weeks, what about if an Airport runway costs more than budgeted, or if groundwater is contaminated by construction? These questions must be answered before a response can be formulated.

How big is the bet we are taking? For large oil companies, the risk of a refinery shutdown running over budget is, at most, an inconvenience (though possibly the person who ran it might have more severe career consequences). The risk of a major mishap during a shutdown is at its worst for a single plant operation. Here the risk could be bankruptcy. The specific parameters of a shutdown (when, how long, how much) might be one of the more risky business decisions that a company can make

### Big Source of Risk

One of the biggest sources of financial and duration risks after you manage resources is added work and emergent work. An added job is a work order handed to the shutdown manager or planner after the work list is closed. Emergent work is discovered after the shutdown commences, usually during disassembly.

Some experts note that work scope might increase by as much as 15 to 40% *after the shutdown has started.*

Other sources of risk include late changes to shutdown, essential staffing being unavailable, delays in delivery of critical spares (particularly sole source), labor action or informational pickets, items to be delivered during the shutdown, weather, accidents, sabotage, design errors and omissions, first-time jobs, misunderstandings, poor estimates, lack of skills, poor communications, contractor financial problems, etc.

Worth an additional mention is the need for specialized skill sets. The availability of these scarce skill sets in your region might vary from week to week depending on what other shutdowns and construction is going on. These specialized skills must be more closely monitored than other skills and back-up plans must be made if adequate hours are not available.

## Risk Management: Steps to take

Three basic steps to manage risk have to be built into the project plan and one step must be added to the active project management process. The three steps to be completed, as part of the planning process are **risk identification, risk quantification and risk response.** The first two, identification and quantification, are sometimes grouped together under Risk Analysis or Risk Assessment. With these three steps in place, the last step (during the project execution) is simply ongoing response to the actual risk conditions on the ground and implementation of the appropriate risk responses.

- **Risk identification:** Is there any risk here? What is the likelihood of the risky event happening? Address both internal (under the team's control) and external (in the outside world) risks. Consider using the Risk Identification process to also look for areas of opportunity in addition to areas of threat.

- **Risk quantification:** How much money will the event cost? How much time will it delay the completion? How many people will get hurt and how hurt will they get?

Opportunities and threats can interact in unanticipated ways and a single breakdown in the schedule can spawn a cascade of problems. One person might suffer from the solution to another's problem. In other words a solution to my problem might cause a risk to you.

- **Risk response:** What response? How costly is it to respond? How likely is it that the response will eliminate the risk? Can the risk be eliminated? Can the risk be transferred to someone else (by using fixed price contracts or buying insurance)? Does the response introduce any unanticipated risk?

**Once we are underway:**

- **Risk vigilance:** How do we organize our team so that when a risk becomes apparent we find out quickly enough that we have time to respond? In addition to vigilance this aspect includes responding to changes in the character of the risks over the life span of the project.

For example a turbine outage in a power plant might take three years to plan. In the early 2001 and 2002 the whole structure of the California Electricity market changed. Any assumptions about risk made in 1998 when the outage was first planned would have to be rethought in light of the dramatically changed marketplace.

With these three steps in place the last step, vigilance (during the project execution) is simply recognition of the risk conditions, the ongoing response to the risk conditions on the ground, and implementation of the appropriate risk responses. Vigilance requires an appropriate trigger. The reasons for our vigilance have to be defined.

## Risk analysis is a team exercise

Conduct a "what if" game with the shutdown planning team looking for events that could result in problems for the shutdown. The team should be people from different backgrounds. Risk is invisible unless the person recognizes that there is a problem. A group with different backgrounds is more likely to see a wider range of potential risks then a homogenous group. In some plants the shutdown managers should bring in outsiders who are experts in some specialized areas and can see risks in those areas.

The shutdown manager must be vigilant for three kinds of risks:

- Risk to the duration of the shutdown
- Risk to the cost of the shutdown
- Risk to the workers, the public, or the environment

These risks could be in the form of disappearing resources, overly optimistic planning, unforeseen circumstances, inadequate engineering, ad hoc modifications, etc.

**Process:**

- Develop checklists of risk elements or dig up the checklist from the last shutdown for modification and improvement. There might be some new risks or some old ones that have plagued you in the past might be resolved, but for the same kind of outage the same risks should apply.
- Interview other stakeholders to identify risks for this shutdown that are outside the immediate project scope. You may find a major risk category that isn't even being considered by your team because it is outside their purview (such as impact on other operations that are not being shut down).
- For each risk identified, decide whether the probability of the risk happening is high, medium, or low.
- For each risk, decide whether the potential impact of the risk happening is high, medium or low. The relative impact is measured in delays in completion, a potential budget shortfall, or problems with safety or environmental compliance.
- Prepare a response if indicated below

## One example

In an electric utility one of the greatest risks is a problem with the cranes during a critical lift. These problems are in several categories. One of these risks is inadequate lifting capacity given specialized lifting parameters (weight, distance from the center of gravity, obstruction, etc. This utility in the Midwestern USA brings in the crane contractor to look over the shutdown package, analyze the lifting parameters and choose the equipment himself or herself. The utility felt the contractor might make different (and better) choices of equipment.

Another risk is that the team may be ready for the placement of one of the critical components in the entire shutdown and the crane breaks down? No crane suitable for the lift was rented for another part of the shutdown, and after much discussion it was decided to rent a back-up crane for critical lifts. The team wanted to insure that the project could continue even if the main crane broke down. Of course there are other ways to mitigate this risk. For example this risk could be dealt with through clauses in the crane rental contract ($100,000 penalty if the crane breaks down in week three, if it cannot be fixed or replaced within 8 hours).

This matrix shows the level of response, given the probability of the risk happening and the impact on the shutdown if the risk does take place.

| | | |
|---|---|---|
| ***High Probability***<br>***High Impact***<br>Response:<br>Full court press-Full Plan | ***High Probability***<br>***Medium Impact***<br>Response:<br>Prepare plan | ***High Probability***<br>***Low Impact***<br>Response:<br>If you want to |
| ***Medium Probability***<br>***High Impact***<br>Response:<br>Full Plan | ***Medium Probability***<br>***Medium Impact***<br>Response:<br>Prepare basic Plan | ***Medium Probability***<br>***Low Impact***<br>Response:<br>Waste of Time |
| ***Low Probability***<br>***High Impact***<br>Response:<br>Prepare Plan | ***Low Probability***<br>***Medium Impact***<br>Response:<br>Prepare minimum plan | ***Low Probability***<br>***Low Impact***<br>Response:<br>Waste of time |

**Risk Matrix shows probability of risk, impact of risk and response**

As each risk is evaluated, a management strategy is adopted for that risk or class of risks.

## Risk recognition and timing

The best time to design the system to manage the risk is usually early in the process. Many risks can be anticipated during the planning process and processes can be put in place to manage them at that time. Much of the planner's job revolves around recognition (identification) of areas where there is a probability that the job will go wrong. An experienced planner can evaluate the risks (quantification) and decide the level of reaction (response).

Risks like delayed delivery of critical spares are managed by tracking each step of the supply process including the verification of engineering completion, verification of commencement of manufacture, etc. until delivery of the parts. Ultimately, most risk management must be part of the individual job plan.

Other risks that emerge during execution include accidents, damage, and miscalculations. Good practices, good planning, and good management minimize these risks.

For example it is good practice (and required by law in the US) to inspect slings, chains, and other lifting gear every day. This practice minimizes the probability of a failure. Another practice is encompassed by good rigging techniques that examine the center of gravity, weight, and material being lifted, and rigs each lift properly. A third practice is the clearing of lift paths so that if the lifting gear does fail, no one will get hurt.

Each of these practices mitigates a specific element of the overall the risk but doesn't eliminate it. With the lowered level of risk, the shutdown owner can, in good conscience, accept the small probability of a failure. If the shutdown can be done efficiently without lifts at all then the risk has been eliminated. The rule of risk management is; if it is possible to eliminate the condition then the related risk is also eliminated. This approach applies particularly to safety and environmental hazards. When you eliminate a risk, of course, be sure you are not introducing a worse risk.

What if the consequences of the part falling are truly catastrophic? What if the lift involves a giant tank of poisonous gas or a nuclear core? I'll bet that the lift planner will go through additional steps to lower the likelihood of failure 100 more times!

## New category of risk and response

In the US after 9/11 the government created a threat level color system. While the system might be redesigned, managing external security threats has now been added to the shutdown manager's standard risk menu. Unfortunately plants that require shutdowns such as power plants, oil refineries and chemical processing plants are also thought to be potential targets.

When the threat level goes up, new security and new access rules are added that could significantly slow down a project. During one elevated threat period, contractors were slowed down by well over an hour a day in entering an airport property to work on a major shutdown project. Every tool, pouch, and box was examined. We suggest that when you are in an area that recognizes threat levels, appropriate language be written into the contracts to protect you from extra charges.

Shutsown personnel present security issues that span from petty theft to major organized grand larceny.

In all cases security is a major consideration for shutdowns and hundreds of people to a site with potential valubles and you have a risky mix. All contractor employees should be known and photographed. Plant personnel should be notified to maintain a high level of alertness.

### There are three choices in management of risk.

In Summery:

- *Avoid the risk*- Much of the work in phase two with schedules, labor load balancing, and scheduling, special equipment should be designed to avoid risk. In other circumstances, the risk can be avoided by changing techniques (build something on the ground rather than in the air, have work de-energized, work cool, etc.) or by reengineering. Not having a dangerous chemical on the site (or having minute amounts) is the best way to manage the risk from that chemical! If you can't eliminate the risk the next step is to mitigate it.
- *Mitigate the risk*- Mitigate to either reduce the probability of the risk happening (using existing technology instead of new technology) or to reduce the financial impact of the risk (buying fire insurance) or some combination of such techniques. Reducing the amount of a dangerous raw material is an effective way to mitigate the risk. Flammable chemicals that are not on the site can't burn! PPE mitigates but does not eliminate the risk.
- *Accept the risk*- You decide that the risk won't happen (like a meteor crashing through the cooling tower). Or you just accept that if X happens the project will take an extra month. If you chose to ignore risks, or chose not to think much about potential risks ahead of time, then you are accepting what might happen. Irresponsible shutdown managers manage a risk on a wing and a prayer.

# CHAPTER 12

# ENVIROMENT, HEALTH, AND SAFETY

**What is a Hazard?** Hazards are one class of risks. Hazards have four forms (according to Lenahan):

These four forms include the damage to the health of an employee, the contractor or the public. This danger is from exposure to a chemical agent, dust, noise, sound, or other byproduct of the shutdown.

A second hazard is to life and limb, which is slightly different from health (although no one would argue that falling is bad for one's health). The danger includes falling, slipping, being hit by falling objects, being cut, and being trapped or crushed.

The third hazard is to property. This hazard includes accidents or planned events that damage buildings, processes, machinery, or property of any kind, including damage to a neighbor's property.

The final form of hazard is to the environment. This danger includes releases of hazardous or toxic materials into the air, water, or ground.

**Safety thrives where the Policy Team requires safety be discussed at every level and at every step in the shutdown.** Safety has to be planned and structured into the shutdown itself at the time of job planning. Safety cannot be added on afterwards, it is like quality, in the sense that if you focus on catching mistakes you are too late. Safety is a class of risk, in one chemical plant all safety activities come under the Risk Management Department.

## A Safety Plan and an environmental protection plan is the beginning

Any shutdown planning function should develop a safety plan and an environmental protection plan. Most of the elements can be borrowed from prior shutdowns. These plans inform all the employees and visitors (including contractor employees) about the procedures to follow for any large hazard such as a fire, gas discharge, explosion, etc. The plans include assembly points, hazard communication bulletins, fire fighting rules, locations and use of first aid stations, safety stations for eye washes, showers and decontamination, and injury evacuation procedures. The plan would provide for alerting all personnel when hazardous actions are scheduled.

## Shutdowns are intrinsically more dangerous (to people, property and environment)

Shutdown work is inherently more dangerous than routine maintenance work for a variety of reasons.

- There are new people in the plant (contractors)
- The work scheduled is not normally done (tank entry)
- More of the work is at a height than with normal maintenance
- The work is larger work (construction)
- Equipment is open and disassembled (the hazard is normally inside so that operations workers on the outside are protected)
- The work site is not stable, and new hazards are being introduced every day
- There are induced hazards (cutting, burning, open lines, and tank entry)
- There is pressure to complete the work on time (the company is losing money every day)
- The heavy overtime causes fatigue
- Many days are worked without time off
- Work continues around the clock
- There is extreme danger right after start-up due to errors in design or execution of modifications and repairs

## OSHA

In the USA the Government at both Federal and State levels is closely involved with industrial hygiene and safety. The Federal agency that is charged with this activity is OSHA (Occupational Safety and Health Administration). In 1970, President Nixon signed The Occupational Safety and Health Act of 1970. The OSH Act charged OSHA with assuring safe and healthful conditions for working men and women. From its earliest days, OSHA was a small agency with a big mission. When the agency opened for business in1971, OSHA covered 56 million workers at 3.5 million workplaces. Thirty-three years later in 2003, 105 million private-sector workers and employers at 6.9 million sites look to OSHA on workplace safety and health issues. (Adapted from the OSHA Website.)

Each State also has an occupational safety and health department, whose regulations cover all work place hazards but are particularly focused on Lockout/Tag out, confined space entry, hazard communications, lift safety, fall safety, and safe electrical practices. Specific industries such as mining, merchant shipping and trucking have agencies focused on the unique health and safety issues for their workers.

OSHA and these specialized safety and health agencies conduct surprise inspections and submit citations (for violations), and if the hazard is not repaired they follow up with fines for infractions of the regulations. If there is a serious accident or fatality, the agency may also do an investigation.

American industry was not in favor of this Federal intervention at the time but workplaces have since become significantly safer. Most organizations follow the OSHA rules and are much safer to work in than they were 25 or 30 years ago. Recently, the agency has been trying to move toward a cooperative model of "safety consultant" and away from the "safety cop" model of the past.

It is particularly difficult to follow some OSHA rules during a typical outage. This difficulty is one of the reasons that shutdowns are more dangerous per hour than routine maintenance work. There seem to be hazards everywhere (and sometimes there are).

For example, if we introduce a fall hazard during routine maintenance we would expect a firm barrier or barricade to make it difficult to get near the hazard. While the rules are the same during a shutdown, caution tape might be acceptable because the hazard will have a short duration. The rule remains the same and to reduce the risks the more stringent regulation must be followed.

## EPA

In the US the agency tasked with protection of the environment is the Environmental Protection Agency or EPA. The U.S. Environmental

Protection Agency was established in 1970 to consolidate in one agency a variety of federal research, monitoring, standard-setting, and enforcement activities to ensure environmental protection. EPA's mission is to protect human health and to safeguard the natural environment--air, water, and land--upon which life depends. When then President Nixon introduced the concept of the EPA in a speech in 1970 he said, "A far more effective approach to pollution control would:

- Identify pollutants.
- Trace them through the entire ecological chain, observing and recording changes in form as they occur.
- Determine the total exposure of man and his environment.
- Examine interactions among forms of pollution.
- Identify where in the ecological chain interdiction would be most appropriate.

In organizational terms, this edict required pulling together into one agency a variety of research, monitoring, standard-setting, and enforcement activities formerly scattered through several departments and agencies. It also required that the new agency include sufficient support elements--in research and in aids to State and local anti-pollution programs, for example--to give it the needed strength and potential for carrying out its mission. The new agency would also, of course, draw upon the results of research conducted by other agencies." (Material adapted from EPA History form www.EPA.gov).

For our purpose during shutdowns, the EPA is the agency that regulates release of gases and smoke into the atmosphere, liquids into the waterways, and waste onto the ground. The agency is also tasked with the disposal of toxic materials and wastes. In the US a firm is responsible for any toxic materials generated forever (the phrase is used from cradle to grave to describe the responsibility).

## Specific Safety Risks to manage

There are several risk classes to manage. The question to ask is, have we tried to eliminate or somehow manage this Safety Risk? The following table is designed to aid in this endeavor:

| Action | Safety Risk Check Sheet |
|---|---|
| | • Airborne contaminants including asbestos, lead dust (from paint removal), chemical fumes (from painting, stripping, open lines, etc) |
| | • Falls from heights. |
| | • Falling objects |
| | • Eye damage (particle, chemical, flash) |
| | • Entrapment and crushing |
| | • Slipping and tripping (falls at the same level) |
| | • Chemical ingestion, skin exposure, breathing |
| | • Fire |
| | • Electrocution |
| | • Explosion |
| | • Asphyxiation |
| | • Temperature stress |

Unlike other risks, safety risks can have a massive effect on people we know and work with. But at its core, these risks can be handled just like any other risk and the process is the same as with other risks.

1. First we have to identify that there is a risk. This action is sometimes tough because the identifying the hazard might take detailed knowledge of chemistry (when you mix A and B there will be an explosion), structural engineering (at 400° C this pipe will expand enough to knock off this T)
2. How much or how bad is this hazard. We must quantify the hazard in terms of money, lives, and environmental impacts. A vapor cloud of poisonous gas that could kill 22 workers is bad enough. In Bhopal, the houses were built right up to the fence. Officially 2153 people were killed (some people think that up to 10,000 people actually died from the poisonous gas cloud) and 200,000 were injured.
3. The last step in the process is to design a response. The possibilities here include the same choices we have for other classes of risks.

✓ Eliminate the hazard. What is not stored is not a hazard. The best choice is to eliminate the long climb, the tank entry, or other hazardous activity if possible. In Bhopal, use of the chemical MIC was an intermediate step in the process to make insecticide and according to a safety specialist it did not even need to be stocked at all (or in just small quantities). No MIC, no Bhopal and no deaths and injuries.

✓ We mitigate risks every day. When you snap on your seat belt or put on your hard hat you are trying to reduce the effects of a known and present hazard. All PPE attempts to mitigate bodily hazards. Sometimes we mitigate by purchasing insurance, buying a back up, outsourcing the job, or instituting a work around.

✓ Some hazards are either so minor or so improbable that we accept them. We may accept small fires from welding or burning operations, minor spills under a few liters, falls under a meter, and other hazards. The remote probability event is also accepted. Perhaps a 500-year water level, 210 mph hurricane winds, or a meteor strike could happen but probably won't. After we do everything possible we have to accept the level of risk that is left.

4. Be vigilant because once the action starts the fog of battle settles over the shutdown. Hazards are being opened up every hour. Situations change, what was safe is safe no longer. Vigilance, along with good solid job plans is your best defense.

For each of the risks listed above we decide: do we eliminate it, mitigate it or accept it? The driver is the consequences and probabilities balanced against costs. If the consequence is grave and the probability is high then the risk cannot be ignored. As you can see in the chart below it is sometimes impossible to eliminate some risks in a shutdown environment. One important precaution (mitigation) used in many fields is to insure that all workers are free from drugs and alcohol. Many shutdowns mandate random drug testing in their contracts. They also maintain a zero tolerance attitude (if someone is tested positive they are banished from the site). This must be clear and part of the safety training (and repeated regularly).

### Job Safety Analysis (JSA), Job Safety Hazard Analysis (JSHA)

A structured review of the hazards of a particular job is called a JSA (Job Safety Analysis) or in some firms a JSHA. The work content and the working environment are reviewed for hazards. Every hazard listed on the Safety Risk Check Sheet is considered. The JSA process produces a document that outlines the hazards of that particular job. It also outlines what mitigation processes are needed for each hazard. This JSA document is included in the individual job packages.

People knowledgeable both about the job and the potential hazards should do this analysis. In some instances the planner looks at the hazards presented by each job step. In others, a safety specialist reviews the job plans when they are completed. Often the contractor performs the JSA and submits it with the job plan for approval.

Working globally has challenges with regard to safety. The JSA is written in some language. The workers might be familiar with another language. In fact on a shutdown in Africa there were three distinct ethnic and language groups. This variety is a challenge for contractors. They are obligated to communicate the hazards to all employees. The owner is responsible to see that communication is effective.

JSA would identify a hazard. The planner or other expert decides if it is to be eliminated or mitigated.

| **Safety Risk Check Sheet** | **Eliminate** | **Mitigate** |
|---|---|---|
| Airborne contaminates | Eliminate contaminates | Respirators, isolate source of contamination, water curtains |
| Falls from heights | Do work at ground level | Guard rails, fall protection |
| Falling objects | Don't allow work at lower levels | Hard hats, shelters, better scheduling and procedures |
| Eye damage (particle, chemical, flash) | Find another way to work, isolate the work, or use robots? | Safety glasses, face shields, isolate source of insult with tent, barriers, etc. |
| Entrapment and crushing | Don't allow work where people can get caught or crushed | Improve procedures, safety shoes, proper clothing, improved chocking, blocking and rigging. Training |
| Slipping and tripping | | Housekeeping, equipment lay down areas well organized, high visibility paint and tape, good lighting, sand mixed into paint |
| Chemical ingestion, skin exposure, breathing | Eliminate hazardous chemicals, substitute with non-hazardous materials | Proper PPE, better procedures, access and understanding MSDS, reduce the need for hazardous materials, reduce stock levels |
| Fire | Eliminate flammables, (note there will always be a source of ignition) | Improved procedures, watch people, pick less flammable chemicals, better fire fighting, inert gas blankets, PPE |
| Electrocution | Lock-out, do not allow work on or near energized circuits | Safe procedures, training, lock-out, PPE |

| | | |
|---|---|---|
| Explosion | Eliminate one or more of the components of the explosive mixture (noting there will always be a source of ignition) | Safe procedures, safety blanket gas, training |
| Asphyxiation | Do not allow entry into any dangerous areas. | Gas monitors, gas alarms, air packs |
| Temperature stress | Do not schedule shutdowns during cold or hot temperatures. | Issue appropriate clothes, work time and temperature tables, cold vests, adequate cold water, shade, evening work |

An article in PEM magazine on safe shutdowns contained a list of the key considerations.

| Safe Work Process Check list |
|---|
| • Work control includes a permit system (that is enforced from the top all the way on down), and training for all personnel |
| • Safe environment includes accommodation for bad weather built into the plan |
| • Safe workplace includes evaluation of the specific work site for hazards including working at height, working near machinery, flame, heat or activated circuits, and confined space entry |
| • Safe task is a review of the tasks to be done for the particular job and amelioration of the risks such as insistence on breathing apparatus near grinding lead paint operations. |
| • Safe worker is essential in a challenging environment. Many hazards can be managed by safe practices or protective equipment, but the worker's approach is an essential ingredient to a safe shutdown. |
| • Safe materials and substances and the safe use, transportation and storage of the materials. All personnel who come into contact with hazardous materials should completely understand the risks and remedies. |
| • The final consideration is the safe process. Management must attend to and manage the risks to achieve the goals of a safe shutdown. |

## Known Hazards

It is essential that the safety officer and turnaround management be able to identify known hazards to the shutdown crew chiefs and eliminate those possible. Nigel Ellis, an expert in falls and fall protection says, "It is not enough to identify the hazard, you must take all means necessary to protect the workers."

His example was of an open hatchway in an area that is not accessed by plant personnel but will be accessed by shutdown workers. He asked, "Is it enough to put up caution tape?" His answer is "no," the hazard needed to be eliminated with a barrier that requires the worker to unlatch a gate.

There are hundreds of known hazards that might include

- Non-load bearing surfaces
- Open ditches
- Elements that stay (electrically) alive during the shutdown
- Oily or greasy surfaces where people might walk
- Structures and supports known to be damaged by corrosion

**What happens when there is an accident during a shutdown?**

Every plant or factory has a procedure to follow for reporting and investigating accidents. This process should be followed (and enhanced) for the shutdown periods. Forms and training will have to be supplied to all contractors and their workers. It is essential to maintain the integrity of the accident reporting system. Every incident should be recorded and reviewed by a competent safety representative. Once on site, all contractor employees are covered by your safety investigation rules (and the contract should reflect this condition).

## Safety Committee

Every shutdown (and plant for that matter) needs a standing safety committee. The concept and organization of the safety committee is important. The mission of the safety committee is to determine what can be done to prevent any accident from happening in the future, in every other plant. The second mission is to make sure everyone knows what happened in any accident and what the company is going to do to prevent it in the future. They also look into determining the way company policy can be modified to prevent similar accidents in the future, to protect the worker's rights, and to involve more of the organization in the safety program. It is not a passive job; the safety committee members are also safety cheerleaders.

Safety Committee members could include management, engineers, workers from the craft or from others, safety, risk management, supervisors etc. The safety committee should not be too large. Four or five people seem adequate. At least one person from outside the plant should be included (if possible).

Anyone interested in safety during shutdowns and normal plant operations should read the works of Trevor Kletz. Titles are in the resource section. I used some elements of his model for accident investigation from the third edition of his *Learning from Accidents.*

**An accident happens, what do we do?**

1. Get the person first aid or to a hospital (of course).
2. Deal with the existing hazards (like PUT THE FIRE OUT) but disturb the scene as little as possible. You may want experts to look over the area.
3. If possible cordon off the area until the investigation is complete. If possible, photograph everything immediately after the accident.
4. Notify any authorities or insurance companies.
5. Appropriate parties should prepare the accident report and report the accident through channels already established.
6. Instruct someone to draw up a list of everyone who was in the area or has touched that machine or process (both before and during the shutdown) for questioning. Include operators, engineers, maintenance workers, contractor workers, and even janitorial personnel.
7. Rework the shutdown schedule with the accident in mind. Try to work around the area if possible or accelerate the physical investigation so that the area can be cleaned up, and try to get back on schedule. Preserve any physical evidence and if you have space remove it to an area where it can be examined.
8. Conduct interviews with the people listed. Be patient in interviewing. Let people tell the story of what happened at their own pace and let them tell the story their own way. Avoid the look, sound, and feel of any kind of police investigation. One goal is to construct a chronology starting well before the event.
9. Gather all relevant information about the accident. This information should include drawings of the machine, piping or area, witness statements, medical reports on the condition of the people (if this is allowed), outside expert reports (such as metallurgy), diagrams or photos, physical evidence, logs of events preceding the accident, work orders, job packages, company and insurance company investigation reports, and shutdown records for the job.

10. Do not investigate to affix blame. Investigate with the intent of finding out what happened. Blame will shut people down because they might want to avoid getting someone into trouble. Trevor Kletz stresses there is enough blame for everyone in a typical accident from the craftsperson that took a short cut, to a designer that made the short cut necessary, to the operations chief that wouldn't make time for proper training, to the CFO that would not spend the extra money for a proper isolation system.
11. Gather all cost data. This data should include repair and downtime costs. Include non-financial data such as accident costs. What were the life costs to any people hurt in the accident?
12. Accidents have causes at many different levels. Close examination might show procedure lapses, poor management decisions, ignorance of parameters, and a final unthinking act. All these mistakes are in the chain of causes for the accident. Use the data to find out what your causes are and how you can remove them. Follow the chain wherever it leads.
13. The focus of the entire investigation is on the question; what can we do to prevent this type of accident in the future.
14. All accidents are caused by many interconnected factors. The factors extend all the way up to the top of the organization (who chose to save money by eliminating ...), to the lowest level (who took a short cut by ...). Some people in the safety field believe that the word accident is inaccurate because it indicates that the event was out of human control. In fact, accidents are preventable and more than that, they are preventable at many levels by many of the people high up on the chain of the organization.

The loss prevention program must have the support of top management. For maximum effectiveness, safety should be a high priority of management. The safety committee should investigate everyone, even top management if they are involved in an accident.

✓ The organization must commit resources in the form of record keeping, appointment of a safety manager (can be part time job with other duties) and formation of a standing safety committee.

✓ What happens to the report and the recommendations contained therein? Organizations that are serious about safety must have a procedure that ensures that recommendations are funded, implemented, and managed for the future.

  In a major accident where a coal mining waste pile slid down to inundate a village the official report noted, "…forty years before it occurred, we have the basic cause of the Aberfan disaster being recognized and warned against…"

✓ The safety or loss prevention program should be applied across the board all the time, not only during shutdowns, to all departments (including sales, administration, etc.). The only special thing about shutdowns is the number of outsiders involved.

✓ The focus of the safety program is people. People are the ones who pay the greatest cost when they are hurt. The program should provide ongoing, effective training as well as written information about any accidents that do happen (as is always done in the airlines industry).

# PROJECT MANANGEMENT TECHNIQUES

# CHAPTER 13
# HISTORY OF PROJECT MANAGEMENT TECHNIQUES

There have been great and wonderful projects in the history of the world. Unfortunately we don't have a good picture of how they managed the building of the Pyramids, the Great Wall of China, or the public works of the Romans. We do have some idea of how great projects of the $19^{th}$ and $20^{th}$ centuries were managed.

In the $19^{th}$ century the great project in the USA was the building of the transcontinental railroad. This great feat was handled as a whole series of interconnected projects (literally) and had many crews working in different locations at the same time.

## Great Projects of the $20^{th}$ Century

The $20^{th}$ century inaugurated projects that were done using modern project management techniques.

## Hoover Dam

The Hoover Dam was constructed on the Colorado River from 1931 to 1936 by a consortium of six major construction companies. The dam was intended to protect people downstream from periodic flood ing, help establish a flourishing local agricultural industry, supply water to the increasingly thirsty state of California, and supply electricity. Ultimately it was the electricity that was to pay for the venture.

The US congress budgeted $175 million in 1928. By the time construction started the depression was in full swing. Like many large projects it was made up of 3 major sub-projects and 100 minor projects. Building of the dam took place in a remote area, 30 miles from Las Vegas, under adverse conditions. Summer temperature frequently exceeded 125 degrees F.

1. In the first part of the project the contractor had to dig four 56-foot diameter tunnels to divert the Colorado River, and line them with 3 feet of concrete. The tunnels had to carry the highest probable flow of the river, and they averaged 4000 feet long. The workers drilled, blasted, and removed 2,000,000 cubic yards of rock to form the tunnels. To create a work site for the dam, they built 2 cofferdams so that the dam site including the tunnels could be dewatered and worked.

2. The next stage was to excavate and pour the concrete for the dam itself. The excavation itself required removal of 7,000,000 cubic yards of rock. The team poured 3,400,000 cubic yards of concrete. Technology was invented and developed to cool the concrete with refrigerated water to reduce cracks, which often occur in concrete as it cools. Until that time, dams were poured and just allowed to cool and the crews then grouted the cracks to seal them. Because of the amount of concrete, the Hoover dam would have taken over 150 years to cool completely. The bulk of the drilling was completed before the concrete was poured, so several air compressors were converted to refrigeration plants (they very cleverly chose compressors that could easily be converted) to supply cool water that was piped through the concrete in 2-inch pipes.

3. The third project was installation of 15 generator units with capacities averaging 115,000 HP.

Some of the smaller projects for the dam included building a city for 5000 people (Boulder, NV), a concrete plant, several miles of railroad (some blasted out of the sides of the canyon), miles of highway, and cable cranes (the largest of which could lift and place 150 ton loads), spanning the Black Canyon.

Extensive planning and scheduling was required to manage the six companies, not to mention 5200 workers and 100 sub-projects.

" Despite the mammoth scope of the project and the multiple numbers of independent companies involved, the active management of the project was left in the hands of four people. Henry J. Kaiser was the Chairman, Charles A. Shea was the Director of Construction, Felix Kahn was Facilities Manager in charge of feeding, housing, and transportation, and S.D. Bechtel directed the purchasing, auditing, and warehouse activities. Underneath Charles A. Shea, Francis T. Crowe was the General Superintendent of Construction. He acted as the point man between the Board of Directors and the operations personnel. It was Shea's responsibility to carry out the Hoover Dam contract on time and within budget. Shea's immediate staff was functionally separated by their specialties such as Traffic Managers, Tunnel Superintendents, Mechanics, Electricians, and Motive Equipment Operators." Compressed Air Magazine (the house mag/azine of Ingersol Rand) wrote a series or articles about building the Hoover Dam in issues from 1931 to 1936.

The Hoover dam was the most advanced and intensive form of project management up until that time. Engineers used Gantt charts to follow the progress of the project. The Hoover Dam is still one of the highest gravity dams in the USA and produces 4 billion kilowatts-hours of electricity annually. The project came in under budget and ahead of schedule. The electricity revenue paid off the 50-year bonds that were sold to fund the dam, in 1986.

## Manhattan Project

In the period between 1942 and 1945 the Office of Scientific Research and Development (ORSD) of the US government undertook the largest R & D project ever, to design and build the atomic bomb. Scientists such as Albert Einstein wrote to and helped convince President Roosevelt to undertake the project (that the scientists and leaders of the day thought actually had little probability of success). The Manhattan project was conceived by leading physicists of the day. They knew that Nazi Germany was working on the same problem and was quite a bit ahead in the science. The leaders knew that the winner of the race to design and build the atomic bomb would win the war.

The project was instituted under the name Manhattan District (to fool spies for the enemy about its real mission). It utilized extensive manual planning and scheduling techniques. At its peak it employed 125,000 workers in Universities and private companies throughout the country. Up to that point it was unlike any other project in size, scope, number of unknowns, and urgency. Basic science for the project was going on while plants were being built (which could be considered the most extreme form of fast tracking). The government funded several alternatives at the same time hoping one would succeed.

The bulk of the research and engineering took place in Chicago (where the first atomic reactor was built under the sports stadium of the University of Chicago), New York City (headquarters for the project engineering firm), and Los Alamos (bomb research). Production took place in two remote facilities. One facility was in Oak Ridge, TN (the location of the U238-U235 separation facility) and the other in Hanford, WA (Plutonium production and separation).

The Oak Ridge main building was 4 stories high and had 44 acres under roof. 20,000 people were involved in the construction of the site. At the time it was the largest building under roof. Across the country in Hanford Washington it took 11,000 pieces of construction equipment. Some 25,000,000 cubic yards of material were excavated for the 1100 buildings at Hanford. At its peak, 45,000 construction workers were employed.

All this was done with extensive secrecy. During the war many materials were scarce. The Manhattan District got an AAA rating and (eventually) a higher X rating (highest rating) to procure materials.

In one story they needed tons of copper for a giant magnet (one of the more promising ways to separate U238 from U235) for an experimental cyclotron. Since copper was in very short supply in wartime, they cast around for a substitute that had very low electrical resistance. The scientists settled on using silver. The only repository of tons of silver was the US treasury in Fort Knox. To keep it secret, Fort Knox lent the silver to the project to be returned after the war (thus there was no need for an act of Congress which would risk alerting the enemy of the strategy). Materials and people were regularly hijacked from other wartime projects.

Large industrial organizations like E. I. Du Pont, Union Carbide, Chrysler, Allis-Chalmers, and others signed up before they found out what they had to do (due to the extreme secrecy of the project). In one example, Allis-Chalmers was given the job of building thousands of compressors that had to operate at supersonic speeds, did not leak (even a little), with no lubricants, containing a corrosive gas (Uranium Hexafloride) that dissolved every material except nickel.

Patriotism motivated these companies. Dupont reluctantly agreed to run the plutonium separation plant under one proviso; that they would be reimbursed their costs plus 1 dollar and no more.

All the project management techniques available at that time were used, particularly fast tracking, where activities normally done in sequence (like design before building) were done in parallel. The budget was $2 billion. The project was successful and tested the first atomic bomb in July 1945.

## Gantt: Modern Project Management started in Philadelphia

The history of formal project management is only a hundred years old and starts where the USA was born, in Philadelphia, located almost midway between New York and Washington, DC.

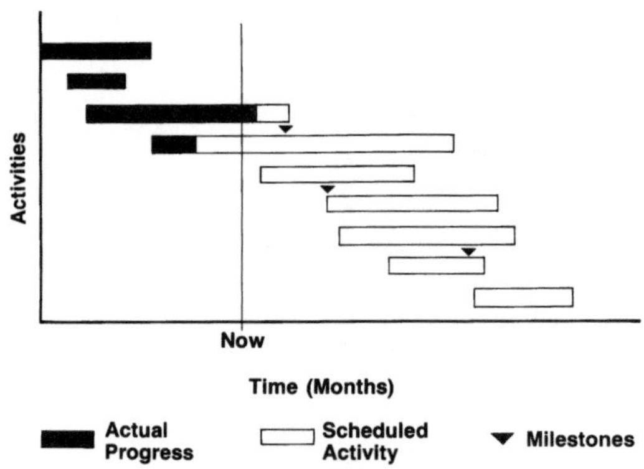


**Image from Defense Systems Management College, 1990 Scheduling Guide**

Harry Laurence Gantt (1861-1919) was a mechanical engineer who specialized in industrial management. He was one of the early leaders in scientific management and worked with Fredrick W Taylor (credited as the father of scientific management).

While working at the Frankford Arsenal in Philadelphia under the command of Colonel George Montgomery (who was in charge of the Arsenal from 1909 to 1918), Harry L Gantt developed a systematic technique for tracking and scheduling armament production. Developed during the outbreak of World War I in 1914, the Gantt chart is one of the oldest planning tools available to maintenance managers.

Gantt was not trying to design a project management method. He wanted to display the estimated production of artillery shells by date versus actual production. His subsequent work before his death in 1919 was in using his advanced charting to display, for all levels of workers and managers, what was really going on in the production areas.

The Gantt chart of the construction project at Reagan National Airport in Washington, DC, shown below was developed using Gantt chart software.

Useful charting software is available on http://www.kidasa.com/

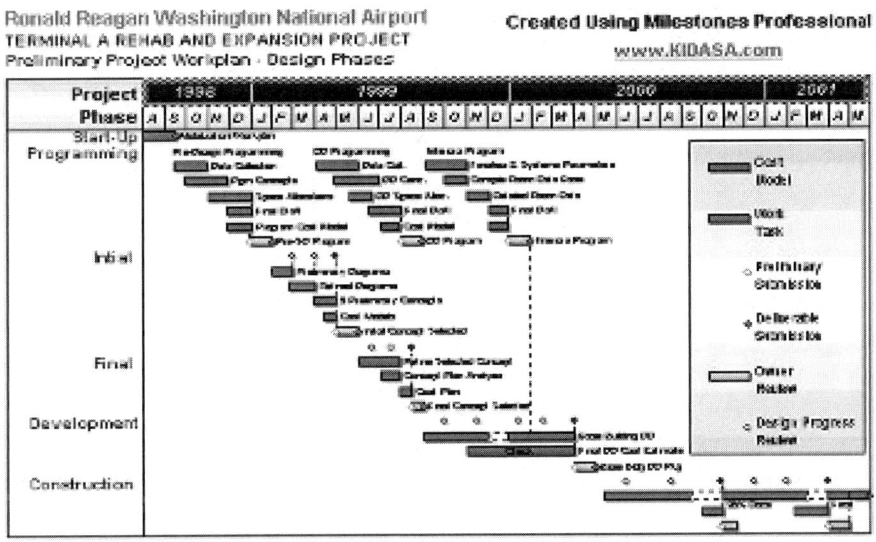


The Gantt chart has the advantage of clearly showing the duration of each activity and relating it to dates on the calendar. It has the disadvantage (in its original form) of not showing the relationship between activities.

Gantt charts are project-planning tools that can be used to represent the timing of tasks required to complete a project. Gantt charts are simple to understand and easy to construct, and they are used by most project managers for most projects.

In a Gantt chart, each task takes up one row. Dates run along the top in increments relating to the duration of the project such as days, weeks, or months. The expected time for each task is represented by the length of a horizontal bar, whose left end marks the expected beginning of the task and whose right end marks the expected completion date. Tasks may run sequentially, in parallel, or partially overlapping.

As the project progresses, the chart is updated by filling in the bars to a length proportional to the fraction of work that has been accomplished on the task. This way, one can get a quick reading of project progress by maintaining a vertical line through the chart at the current date.

Completed tasks lie to the left of the line and are completely filled in. Current tasks cross the line and are behind schedule if their filled-in section is to the left of the line. Tasks are ahead of schedule if the filled-in section stops to the right of the line. Future tasks lie completely to the right of the line.

### Keep it manageable

In constructing a Gantt chart, keep the tasks to a manageable number (no more than 15 or 20), so that the chart fits on a single page. More complex projects may require subordinate charts. Each subordinate chart might represent a single bar of the original chart. For team projects, it often helps to have an additional column containing numbers or initials that identify who on the team is responsible for the task.

Often the project has important events, which you would like to appear on the project timeline, but which are not tasks. For example, you may wish to highlight when a prototype is complete, or the date of a design review. You enter these data on a Gantt chart as "milestone" events and mark them with a special symbol, often an upside-down triangle.

Computerized Gantt charting software stores more information about tasks, such as the individuals assigned to specific work and notes about the procedures. They also offer the benefit of being easy to change. Charts may be adjusted frequently to reflect the actual status of project tasks as they vary from the original plan. All PMS (Project Management Software) have Gantt chart capabilities. The original chart is called the baseline. All subsequent charts are compared back to the baseline.

Gantt charts are widely used by the construction industry. There is widespread expertise in Gantt charts and they are part of business school curriculums.

To use a free Internet based Gantt charting tool go to http://associate.com/gantt/.

### CPM and PERT

Much of the theoretical underpinning of project management, which was driven by actual projects, happened between 1914 and

1956; the two biggest and most influential projects of the first half of the 20$^{th}$ century were The Hoover Dam and the Manhattan Project. On the surface, however it seemed that not much was happening until computers were applied to the problem. When computers were introduced, the field moved forward like a wildfire.

Project management was one of the early programs designed for the first computers (right after the military applications). For the computers of the day, PM was a killer application. In other words, organizations bought computers to get the advantage of PM.

**UNIVAC I Similar to the one used for CPM at Dupont**

## CPM (Critical Path Method)

In 1956, M. R. Walker of E. I. du Pont de Nemours & Co. and J. E. Kelly of Remington Rand developed a computerized system of managing engineering projects. It was called the Critical Path Method or CPM for short. The first test was made in 1958.

For the information of computer lovers the computer used was the UNIVAC I, which by this time, was well accepted. The first UNIVAC I, delivered 7 years earlier in 1951 to the U.S. Census Bureau, was the first commercial computer to attract widespread public attention. It cost $750,000 at the time and could run 1,905 operations per second and had 1,000 12-digit words of memory. It took up 943 cubic feet.

The project was to manage construction of a chemical plant to produce catalysts, being built by Catalytic Construction.

The first shutdown run with project management software and techniques was in March 1959. The CPM method was applied to maintenance shutdown at the Du Pont works in Louisville, Kentucky. Duration of the planning phase was reduced from 125 to 93 hours.

The ascendancy of both CPM and PERT over the Gantt chart was the identification of a particular (critical) path (a sequence of activities) through the project that takes the longest possible time (there are several definitions of critical path that we will review as we proceed). It was found that managing the Critical Path(s) ensured timely completion of the project.

## Some Limitations of CPM

CPM operates on the assumption that there is a precise known time for each activity in the project. This assumption may not be true in real life. CPM does not incorporate any kind of statistical analysis in determining time estimates.

The second limitation was more important in the past when computers were less powerful. It is difficult to use CPM as controlling device for the simple reason that one must repeat the evaluation of the entire project each time changes are made. It might have taken hours for each run through. CPM was initially developed as a static planning utility and not as a dynamic controlling device. In today's computing environment, even large projects can be run in a few minutes, so this limitation is less important.

A third limitation of both network methods is the difficulty of seeing the project's status at any given time. The network connections (arrows and boxes) are drawn time independent so it is easy to confuse the timing relationships between events. As mentioned, these techniques were conceived of more as planning and sequencing tools, and not management tools. Project managers switched to Gantt charts to manage their projects. A partial solution has been the development of time-based networks where the length of the arrows is related to the duration of the project.

## PERT (Project Evaluation and Review Technique)

At the same time, in 1956, the US Navy began pursuing the development of the Polaris missile. The Polaris missile project coupled design of a ballistic missile with the ability to launch those missiles from a submarine platform. The best platforms of the day were the new nuclear reactor powered submarines. The first one, Nautilus was launched on September 30, 1954 and had been operating under nuclear power for over a year while the George Washington was being designed.

The project was daunting. USN ONR (US Navy Office of Naval Research) was faced with a multi-year project that involved scheduling 250,000 major activities with 250 contractors and 9000 sub-contractors. The consultants Booz, Allen & Hamilton, and Lockheed (the missile contractor), are credited with developing the Project Evaluation and Review Technique (PERT).

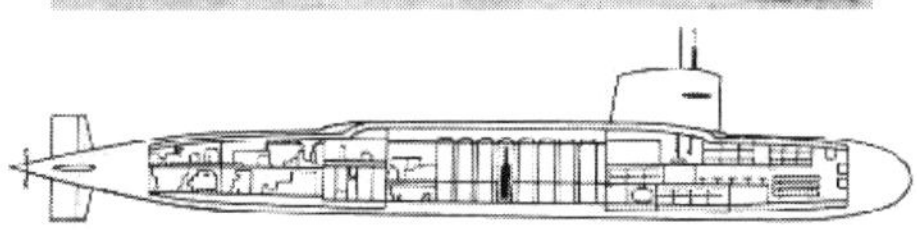

The submarine George Washington was laid down on November 1, 1957, launched on June 9, 1959, and commissioned on Dec. 30, 1959. Commander James B. Osborn was in command of the Blue crew and Commander John L. From, Jr. was in command of the Gold crew. "The George Washington sailed from Groton, Connecticut on June 28, 1960 to Cape Canaveral, Florida, where she loaded two Polaris missiles.

Standing out into the Atlantic Missile Test Range with Rear Admiral W.F. Raborn, head of the Polaris Submarine development program, on board as an observer, she successfully launched the first Polaris missile from a submerged submarine on 20 July 1960. At 12:39 George Washington's commanding officer sent President Eisenhower the message: *POLARIS - FROM OUT OF THE DEEP TO TARGET. PERFECT.* Less than 2 hours later a second missile from the submarine also struck the impact area 1,100 miles down range". (from the open Enclopedia Project in the resource section)

The George Washington was fitted with the Polaris A-1 missile. The Polaris was a breakthrough in size, accuracy and range. "The A-1 Polaris -- solid-fuel, compact at 28 feet long, and bearing a range of 1,200 miles -- was ready for deployment by 1960. The A-2 version with 1,500-mile range, entered service in 1962 followed a year later by the 2,500-mile A-3. All of which could fit in the same launch

tubes". (PBS Nova special on Submarines)
Characteristics of the USS George Washington:

- Displacement 5959 tons surfaced, 6709 tons submerged
- Length 116.3 meters (381.6 feet)
- Beam 10 meters (33 feet)
- Speed 16 knots surfaced, 22 knots submerged
- Missile tubes: 16
- Torpedo tubes: 6 forward, 21-inch
- Complement: 12 officers - 100 enlisted men (each in 2 crews -gold and blue)

NORC computer control console

PERT software was first run on the IBM Naval Ordinance Research Computer (NORC) at Dahlgren, Virginia. For computer buffs the IBM's Naval Ordnance Research Calculator (NORC) was the first supercomputer and the most powerful computer on earth from 1954 to about 1963 (when the twin processor CDC 6600 supplanted it). Built between 1950 and 1954 at Columbia University's Watson Scientific Computing Laboratory, NORC's specifications included:

- Decimal integer and floating-point notation and operation.
- Word size: 16 decimal digits + check digit (64 + 2 bits).
- Clock: 1 μsec.
- 15,000 operations per second with automatic error checking.
- Add time: 15 μsec. Multiply: 31 μsec. Divide: 227 μsec.

PERT expands the CPM method to consider the duration of each activity a statistical function. For each activity three estimates are recorded: the optimistic, most likely, and most pessimistic.

According to Hugh McCullough, the former Polaris project business manager, PERT had a disciplinary effect. He said the Polaris project had a 20,000-event network and the application of PERT to help manage all those activities was very successful. He credits PERT with saving two years in bringing the Polaris Missile to combat readiness.

## Limitations of PERT

The time estimates constitute a major limitation (of PERT, CPM and all Project Management Systems). If the time estimates (in this instance, three estimates for each activity) are unsatisfactory, then the network diagram and the critical path will have little real meaning after the project begins.

The calculation of probabilities under the PERT approach is done on the assumption that independent activities are statistically unrelated to each other. A problem with a single resource (such as an accident) could throw off all activities that use that resource. Activities that are related to each other cannot be expressed by PERT.

PERT suffered from the same problem as CPM in that using PERT for active control of a project required frequent recalculation, which was time consuming and costly. Project Managers would pull out their simple Gantt charts to manage the day-to-day activities.

## CPM won

The simpler CPM method ended up winning the battle for dominance of the project management field. The use of PERT extended dramatically throughout the defense industry after the Polaris success. But by the 1970's, PERT usage declined almost as dramatically. It was thought that PERT was inappropriately applied across the board when the cost to collect and enter the data soon became almost as large as the cost of the project!

The defense industry returned to simpler methods (reference *Scheduling Guide for Program Managers*). Pure PERT charts are rarely used in industry today. Fortunately elements of PERT, CPM and Gantt have been adopted into the Project Management software so they are all in use today.

The CPM/PERT chart or network diagram is sometimes preferred over the Gantt chart, because it clearly illustrates task dependencies. On the other hand, either of the network charts can be much more difficult to interpret, on a day-to-day basis especially on complex projects. Generally project managers use both techniques. The ADM Network chart below is typical in being difficult to interpret and see exactly what is going on.

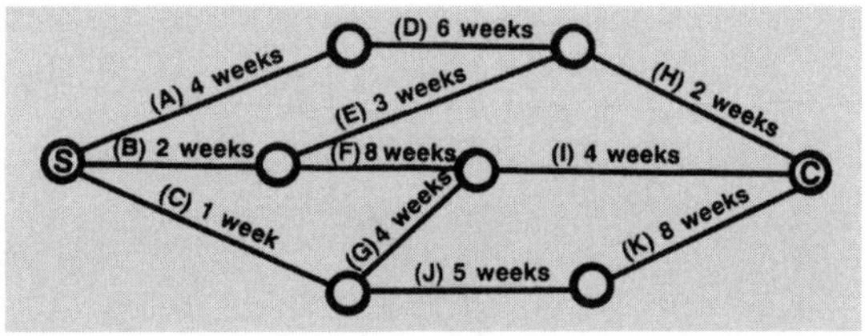


Images from Defense Systems Management College, 1990 Scheduling Guide

Practitioners have adapted the Gantt chart with float and some ability to see dependence on a new hybrid form (for the same network as above):

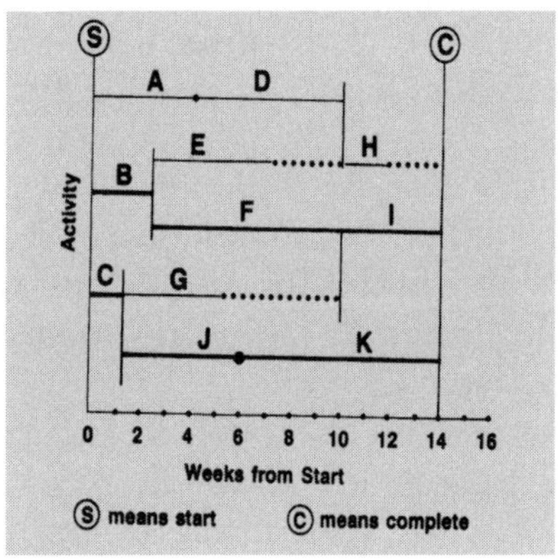


Image from Defense Systems Management College, 1990 Scheduling Guide of the same network as the figure above.

# CHAPTER 14
# KEY CONCEPTS FOR ALL PM TECHNIQUES

**Schedule:** (defined by Webster) *a procedural plan that indicates the time and sequence of each operation.* The key concept in this definition is time and sequence. For our purposes a schedule is a specific kind of plan where time (when something will happen), sequence (how is it related to other activities), and duration (how long will it take), are added.

By transferring all the information (about time, sequence and duration) from the individual job plans to the schedule and manipulating it to optimize duration, resources, or budget, we can create a shutdown master schedule.

Once all the planned job packages are completed, the shutdown planners start to schedule the jobs. There are many of ways to accomplish this scheduling function. For simple projects it is only necessary to draw a chart of the individual jobs along with their durations. In complex projects some computer aid is necessary. The jobs are scheduled in general and then by day or shift. The best way to control any shutdown is to manage what should be happening at the time it should be happening!

**Why go to the trouble of scheduling using Project management?**

1. Collisions of Labor/Material/Tooling/Machine/Order are substantially easier, cheaper, and faster to resolve on paper than in the field. This statement is a golden rule of scheduling. Collisions are uncovered in the planning phase and dealt with in the scheduling phase. They are dealt with again in greater detail in the execution phase. Constant vigilance is necessary to manage these potential collisions. Collisions have a cascading impact. As an activity gets delayed it affects everything in its path, so small

collisions can cause major dislocations (and can cost big money to resolve).

2. Project Management network techniques identify a group (or sequence) of activities within the project whose durations regulate the length of the project. The longest path through the project (that includes these activities) is called the critical path. The breakthrough benefit is that, if you keep the ever-changing critical path on schedule the project will run on schedule. In practice this rule means that you do not need to manage everything, only the vital few things either on or near the critical path. This aspect was a great time management boon to project managers.

3. The second benefit related to the critical path is, if the critical path falls behind schedule you know:

A. The whole project is in trouble and it will not be completed on time.

B. Only an intervention such as increases in the number of people, adding overtime, (or other critical path compression technique – discussed in a later section), will bring the project back on track.

4. We know that time estimates are more likely to err by being too long rather than too short. On the Polaris program, a distribution called the Beta distribution was used, which is not symmetrical. The benefit of including the pessimistic estimate is dramatically increased accuracy of project estimates.

**Activity:** A job or task necessary to complete a project. An activity has certain attributes.

1. Activities have durations
2. Activities consume resources (labor hours, parts, tools, etc.)
3. Activities have clear starting points and ending points

Added into the definition for CPM and PERT (based on the figure that follows)

4. Activity 1 consumes resources A, B, and C and Activity 2 consumes resources A and D.
5. If a successor activity (one that comes afterward) must start in the middle then the first activity must be divided into two.
6. (In the figure below) activity 1 is a predecessor to activity 2

7. Activity 2 is a successor to activity 1

8. Activity 1 must be complete before activity 2 can be started

An activity is the unit of action in project management. This is true whether you are using Gantt charts, CPM or PERT networks, or some combination. In some applications, a single work order is an activity. In others, a work order is made up of multiple activities. Much of the challenge of project management is identifying activities that are manageable (not too long and not too short).

The challenge is to be sure that the activity is not too big to manage. If an activity has 4-hour duration then an intervention can be made within that day or the day after at the latest. If the activity is of 480 hours duration (without clear milestones) then we won't necessarily know it is behind schedule until 60 shifts into the project (which may be too late).

At the other end of the spectrum are the too short activities and the associated paperwork and management effort. By definition, more activities require more paperwork and management time and effort. The happy medium is to choose an activity of sufficient duration to be able to manage without getting buried in paperwork.

### The Key is in the Display of the structure of the Shutdown

While the mathematics of PERT are impressive it is the display of the relationships that is important. A CPM/PERT chart presents a graphic illustration of a project as a network diagram consisting of numbered nodes (either circles or rectangles) representing events, or milestones in the project, linked by labeled vectors (directional lines) representing tasks in the project. The direction of the arrows on the lines indicates the sequence of tasks. By convention, if there are no arrows the network is read left to right.

The tasks that must be completed in sequence are called dependent or serial tasks. Tasks that do not depend on each other can be undertaken simultaneously. These tasks are called parallel or concurrent tasks. Tasks that must be completed in sequence but that don't require resources or completion time are considered to have event dependency. These tasks are represented by dotted lines with arrows and are called dummy activities.

### Arrow Diagramming Method (ADM)

The original method to display the network uses the arrow diagramming method (ADM). PERT charts all used this method and it can still be seen in textbooks on the topic.

The arrows represent the activity and resources, and the circles are the events. The length of the arrow does not denote duration. If the arrow is Bolt Flange then the circle at the end (7) is Flange bolted and at the other end (6) is Flange bolting commencing.

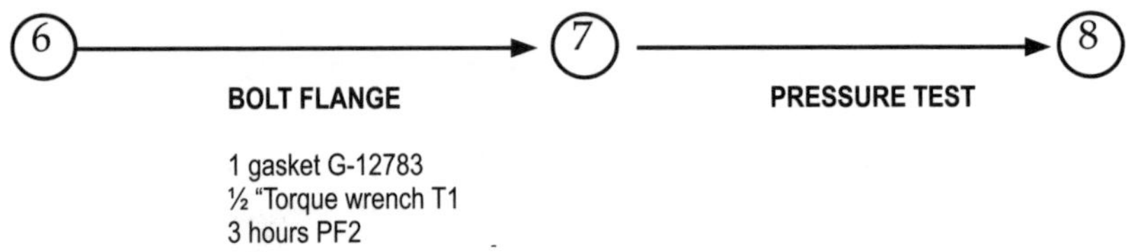


The dependence of a successor activity is show by having them share an event or node. So node 7 is shared by activity Pressure Test. Node 7 is Flange Bolted and Pressure test commence. The Pressure Test Commence is dependent on Bolt Flange completion.

In standard diagrams, some or all the resources are listed under the activity. Each resource has a different code. A journeyman pipe fitter might be PF2 and a torque wrench might be T1. Parts might be coded with their storeroom number.

So to bolt the flange we might see a gasket, torque wrench, and the amount of time and craft. You could read through the network and see most of the important facts. The computer could sum the resource requirements for each type of resource and prepare a report of what is needed when (how many hours or days, etc. after the project started).

### Precedent Diagramming Method (PDM)

In both methods, all information is still displayed and available. Some people prefer one type of display to another. MS Project® and others have adopted PDM so that it is becoming the most common and most taught method.

PDM focuses on the activity or process and the product or event falls into the background. In PDM the activity and resource requirements are in the box and the arrow is only the relationship. As in ADM, the length of the arrow is not related to duration. In fact in PDM, even the arrowhead might be dispensed with.

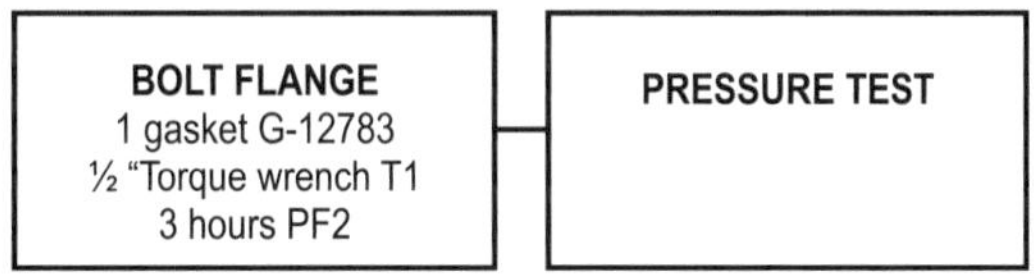


PDM diagrams are read from left to right. The right one is the successor activity to the left one (and cannot be started until the left one is complete) and the left one is the predecessor to the right one.

**Special configurations in network diagrams have special names**

*Burst Point:* When several activities are successors of a single activity. Burst points are typical of the beginning stages of a project. The diagram seems to burst open at this point. All of the successors are dependent on the completion of the predecessor (activity 1).

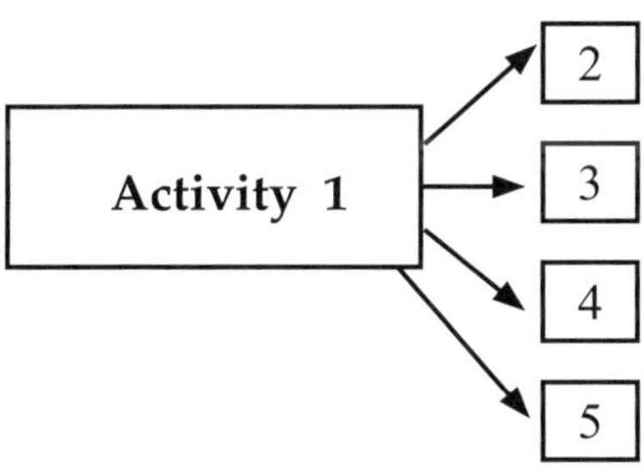


Burst Point (the network opens up

*Sink:* The reverse of the burst point. Several activities come together to a single point. A sink is typical of the end of a project. The successor activity (activity 6) is dependent on the completion of all the predecessor activities.

**Sink**

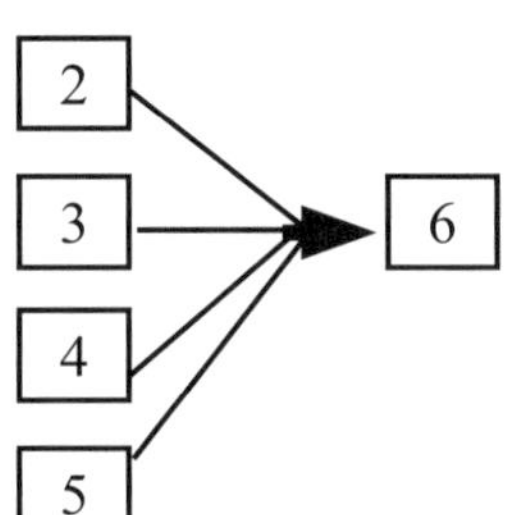

## Milestones

Milestone charts are event or goal oriented. Gantt charts are activity oriented. The milestone concept was developed in Gantt charts (or rather shoe-horned into Gantt charts) and is used in all PM techniques as a way of advertising the progress of the project. A milestone is an activity without duration (another name for an event). It is like a 'milestone' on the side of a road showing how far you have to travel (and how far you have traveled).

Milestone charts are extremely useful for outsiders to measure the progress on the critical elements of a project. Top management might have just milestone charts and milestone meetings. That way important signposts are observed and reported, and the project team is kept accountable. Well-chosen milestones also keep the project team honest because they have high visibility and encourage accountability.

Milestone charts are not particularly useful for determining any other information beyond the project status. Resources and critical paths cannot be seen in milestone charts. Because they are visually simplified these charts are excellent tools for discussions of project timeliness.

Milestones should be chosen for their importance to the project, easy measurability, and intrinsic interest to outsiders. Not every completion of an activity is a milestone. One organization, Landmark Education has a unique way to set up milestones on life projects (as opposed to business projects). They say when designing a project, imagine that today the project is completed. Think back to what was the last big thing that happened to make the completion of the project today, as they say, "a done deal." Continue the process of picking the outstanding measurable events back and back until you reach the beginning of the project. The Landmark milestones might not even be related directly to an activity.

Once set up, these milestones help the project team see what action has to be done to assure the making of the milestone by its due date.

As mentioned earlier, in the 1970s the US military shifted back to Gantt charts with milestones. They developed elaborate milestone nomenclature to display many aspects of their projects. In fact, some of the aspects of CPM and PERT were incorporated into this milestone method. On the following page are some of the conventions used by the US Air Force on their milestone charts for project management.

Standard symbols have been adapted for Air Force milestone schedules. The most common symbols used and their meanings are shown below.

**Basic Symbol** / **Meaning**

Schedule completion

Actual completion

Previous scheduled completion—still in future

Previous scheduled completion—date passed

**Representative Uses** / **Meaning**

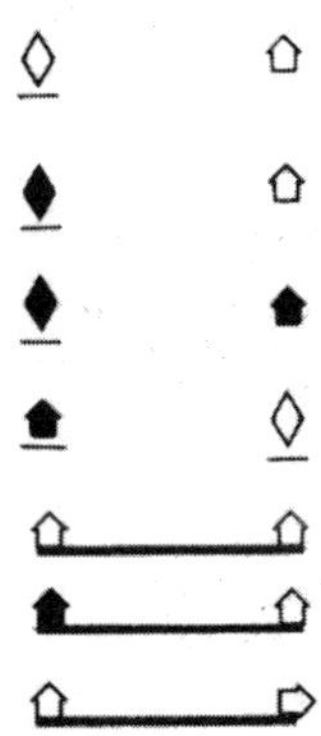

Anticipated slip—rescheduled completion

Actual slip—rescheduled completion

Actual slip—actual completion

Actual completion ahead of schedule

Time span action

Progress along time span

Continuous action

CHAPTER 15

# CRITICAL PATH METHOD (CPM)

The Critical Path is the shortest time through a project network where all activities are completed, or the longest possible path through the project. This concept was an essential breakthrough for project management. It is the defining difference between the CPM, PERT methods, and the Gantt chart. If you manage the critical path, the near critical path, and the keep an eye on upcoming issues along the critical path, then the project will run as efficiently as possible. Delays in any critical path activity will result in the project being late.

"The Key Concept used by CPM/PERT is that a small set of activities, which make up the longest path through the activity network, control the entire project. If these "critical" activities could be identified and assigned to responsible persons, management resources could be optimally used by concentrating on the few activities that determine the fate of the entire project" Ajit Mani, founder of Intervention, an India-based consultancy.

**Developing a Chart for a Shutdown (assuming you have PMS)**

Start with the approved work list for the shutdown.

**Pass 1:** Enter the approved work list into your PM software by work order number.

**Pass 2:** Add estimated duration (hours, days, weeks) for each job based on the optimum crew size.

**Pass 3:** Add the relationships between jobs called dependencies (what activity (s) must be done before each of the activities)

**Pass 4:** Add resources for each work list item including labor (duration times crew size), materials, tools, heavy equipment, etc.

**Pass 5:** Either the system can be set to give you the start date or you can set the start date. When the actual start-up date is known, re-run the chart from that point. The system can also re-run the chart to end on the day given.

## Simple way to determine dependencies

One of the toughest steps is to determine dependencies. It can be overwhelming to look at the entire work list and try to determine what is first, second, etc. A simple way to do this (recommended by Mike Brown of NSI in the resource section) is to ask a simple question for each activity on the list, starting with the first activity on the list:

***What activity immediately precedes this activity?***

Don't worry about what is first in the whole project but just what has to be done immediately before the item you are reviewing. When you have applied this simple question to all the items on the list you can either make a manual drawing of the network or enter the data into the software. Once in the software, this list of dependencies is used to create the entire network (answering this question will allow the software to generate the network diagram no matter how complex).

## Simplified Network to show successors and predecessors

To simplify the activities of building a pipe section (given that the pipe, fittings and drawings are available) the following sequence might be useful.

| Activity | Activity description |
|---|---|
| 1. | Measure pipe |
| 2. | Cut pipe |
| 3. | Thread pipe |
| 4. | Assemble pipe and fittings |

In the old PERT charts, events would be more important than activities so the simple example above would be written: Pipe measured, Pipe cut, Pipe threaded and Pipe Fittings and Assembly complete. For each activity, the activity immediately preceding is shown.

| Answer to the question: What activity immediately precedes this activity? | The series of mini-diagrams depicts the precedence for each activity: |
|---|---|
| Activity #1 there is no activity immediately preceding this activity | 1 |
| For Activity #2 the #1 is the activity immediately preceding this activity | 1 → 2 |
| For Activity #3 the #2 is the activity immediately preceding this activity | 2 → 3 |
| For Activity #4 the #3 is the activity immediately preceding this activity | 3 → 4 |

From the simple statement network diagrams, a complex network can be constructed

The above 4 statements are linked to get the following simple network:

1 → 2 → 3 → 4

## Types of dependency Relationships

To complicate matters in the real world of maintenance shutdowns there are three types of dependency relationships. In the major PM software activities, dependency can be defined as:

1. Finish to start - This is the most common. All examples use this dependency relationship. The framing cannot start until the foundation is cured.

2. Finish-to-Finish - Two activities must end at the same time. Ideally when a new machine is installed, the operator training should conclude when the installation is complete so that the machine can be used immediately.

3. Start to start - Collecting information for the shutdown report starts when the shutdown project is initiated. Those activities should start together.

### Dummy Activities

In some ways computers are not too bright. Every activity has to have a unique starting point and a unique ending point. This starting-ending point is the unique address of the activity. Dummy Activities were invented to satisfy certain logical problems that came up. The dummy activity has zero duration and is only used to show a logical relationship in the arrow diagramming method. Dummy activities are used when logical relationships cannot be properly described with regular activity arrows.

### Lag time

Although it might take 2 days to pour a foundation, it takes 7-30 days to cure it to functional strength. The masons are available as a resource to pour other footings or finish stonework but the next activity cannot start on top of the new foundation before the curing is complete. The wait time between these activities is called lag time. This kind of lag time is known as Finish to Start lag time.

If a bridge is being given two coats of paint, the second coat might lag the first one by 12-24 hours. We could paint for a day and start a second crew the next day. This Lag is known as Start to Start.

Finally Lag can be Finish to Finish. For instance, Quality inspections will lag pipe installation so that completion of inspections depends on or Lags the completion of the piping. It sounds complicated but put simply; inspection can follow the completion of each independent section. But the inspection cannot be completed until the whole piping system is completed.

### Dependency mistakes:

- Loop: making 'A' a predecessor of 'B' and 'B' a predecessor of 'A'. Sometimes activities A and B happen a number of times such as when adjusting a process and measuring the output, then readjusting the process based on the measurement. If this sequence is used, the activities are to be listed as many times as necessary to get it right. In large projects, loops are not always visually obvious.

- Dangler: A task without any successors. This condition is very common in shutdowns because many work orders are not really related to anything else. The solution is to connect all the danglers to a dummy task called end of project (which will allow the critical path to be calculated).

- Hanger: Similar to a dangler but usually the result of an oversight. An unintentional break in a network usually caused by a missing activity or a missing logical relationship

- Redundancy: If A is a predecessor to B and B a predecessor to C, and you list A as a predecessor to C, you have redundancy. Only immediate predecessors need to be listed. This mistake is common when answering the question "what activity immediately precedes this activity?" The tendency is to list all the successors to the activity instead of just the immediate successors.

## Float or Slack

For many activities, late starts will not impact the overall project. The term for time before a delayed activity becomes critical is called slack time or Float. Until the float is consumed the late activity is not critical. When the float runs out, non-critical path activities become critical. Another definition of Critical Path is a series of activities without float.

Three forms of float are quantified to allow project managers and planners to assess the impact of delays quickly (a fourth form of float is recognized by its absence- negative float):

- Total Float – is the delay in activity start or the increase in activity duration that can be absorbed without delaying project completion. The formal definition of Total float is Total Float = $\text{Time}_{\text{LFT (Latest Finish Time)}} - \text{Time}_{\text{EFT (Earliest Finish Time).}}$ The latest time puts the activity on the critical path (in other words it has been delayed to the point where it just becomes critical). The earliest time is the sum of the times from the start of the project along that path.
- Free Float – is the delay in activity start or the increase in activity duration that can be absorbed without impinging upon the timely start of any immediate successor activity. Free Float is also defined as EST (Earliest Start Time) of Successor – EFT (Earliest Finish Time) of successor. (This formula separates out float added by the current activity).
- Independent Float – (also called interfering float) is the delay in activity start or the increase in activity duration that can be absorbed without adversely impacting any other predecessor or successor activity.

- Negative float or Negative Slack- The situation in which an activity or a branch is already behind. This condition is common where the network is designed with the shutdown start time and shutdown end time as constraints. Compression techniques have to be applied, or some other factor changed, to bring the shutdown in on time.

Total Float is shared. All the activities on any branch that has float, share that float. In other words, if the float is used up by the first activity it is used up for the whole branch. In fact, if the first activity uses up the float, then the whole branch becomes critical.

**There are four important times for each activity.**

Project management software calculates these times (which is a good thing if the network is of any size) to determine critical path and float. To determine the critical path and the amount of float, both the LST and EST calculations must be made from a forward and then a backward pass through the project.

| | Name | Description | How to calculate |
|---|---|---|---|
| EST | Earliest Start Time | When is the earliest time this activity can begin? | Calculate it on the Forward Pass. It is the same as the LFT (Latest Finish Time) of its predecessor. If it has multiple predecessors then it is equal to the latest of the LFT (Latest Finish Times) of all its predecessors. |
| EFT | Earliest Finish Time | When is the earliest time this activity can end | Calculate on the Forward Pass by adding the EFT (Earliest Finish Time) of the predecessor plus duration of that activity. If it has multiple predecessors then use the latest of the (EFT) Earliest Finish Times of all its predecessors plus the duration. |
| LST | Latest Start Time | When is the latest time this activity can begin without impacting the duration of the project | Calculate it on the Backward Pass starting with the LST (Latest Start Time) of the successor less the duration of the activity. If there are multiple successors take the earliest of the LST (Latest Start Time) less the duration |
| LFT | Latest Finish Time | When is the latest time this activity can end without impacting the duration of the project | Calculate it on the Backward Pass it is the same as the LST (Latest Start Time) of the successor. If there are multiple successors take the earliest of the LST (Latest Start Time). |

Given the Network diagram on the next page, the float and critical path can be easily calculated. First do a forward pass through the network to get the Earliest Start Times (EST). A simple calculation (adding duration) gives the Earliest Finish Time (EFT). A reverse pass gives the Latest Finish Time (LFT) and a simple subtraction gives the Latest Start Time (LST). With these numbers the Critical Path and the Float can be determined.

Some systems draw a big box for each activity with all the information (PDM):

| | |
|---|---|
| **EST** (Early Start Time) | **EFT** (Early Finish Time) |
| **Work Order Number** | Task Duration |
| Task Description | |
| Major Resources | Float |
| **LST** (Latest Start Time) | **LFT** (Latest Finish Time) |

Critical information for any activity

**Simple Boiler Shutdown**

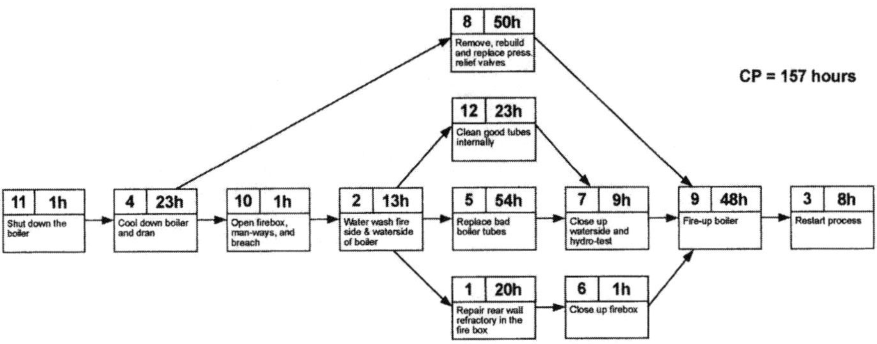


## Simple Shutdown Network

We have decided to shutdown our boiler to make some repairs that cannot be completed while the boiler is still hot. Since the boiler will be taken down, all the processes that require steam will also be shutdown. This would be a good time to review your backlog and schedule any work orders that require the steam to be off. It is important to return the boiler to service as soon as possible since production must be shut down.

The above Figure is the network diagram for this shutdown. It shows the critical path graphically. The following figure shows the detailed calculations to arrive at the float and critical path.

**Forward Pass:**

1. Start with A (activity) -11 and determine the earliest start time (EST), which is zero.
2. Add in the duration for A-11 (1 hour) to get the earliest finish time (EFT), which is 1 hour. If A-11 takes 1 hour then it cannot finish before 1 hour hence the Earliest Finish Time or EFT.
3. Go to next activity A-4 and enter the EST (earliest start time), which is the same as the predecessor's (A-11) earliest finish time or 1.
4. Add the duration of A-4 to the EST of A-4 to get the EFT (1 + 23 = 24). Again if A-4 takes 23 hours then it cannot finish before 24 hours, hence the name Earliest Finish Time or EFT.
5. Step through the network, putting in the EST and EFT for each activity in sequence. When an activity that has multiple predecessors is encountered, such as A-7 or A-9, compare the EFT of all the predecessors and choose the latest EFT. The predecessors dictate when an activity can start. All predecessors have to be complete before the successor can start. That is why the latest of all the EFTs is used.
6. Once the end of the network is reached, the EFT of A-3 is the length of the project.

**The Backward Pass**

1. Start at the end of the network with A-3 and the EFT is also the Latest Finish Time (LFT) or 157.

2. Subtract the duration (8 hours) from the LFT to get 149 as the LST (Latest Start Time) for A-3. This time is the latest that A-3 can start without impacting the duration of the whole project.

3. Proceed to the predecessor activity A-9. The LFT of A-9 is the LST of A-3. Subtract the duration of A-9 to get the LST for A-9. Again, this is the latest time that A-9 can start without impacting the duration of the whole project. Since A9 and A3 are on the critical path, you'll notice that the LST and EST are the same. There is no float on the critical path.

4. If there are multiple predecessors (like A-9), the LST of the successor is the LFT of all the predecessors (A-6, A-7, and A-8).

5. When an activity is reached that has multiple successors (such as A-2 or A-4), compare the LST of all the successors (for A-2, the successors are A-1, A-5 and A-12 with LST= 80, 39, 69 respectively), and choose the earliest or smallest number (here 39) as the LFT.

6. When the beginning of the network is reached (A-11), the LST and the EST should both equal zero. If they do not, there is a mistake in the arithmetic.

Another definition of the Critical Path is the path through the network where the Earliest Start Time and the Latest Start Time is equal. Float is defined as the time between the LFT and the LST so if float is zero then the activity is on the critical path.

| Activity | Duration | Float | Note | EST | EFT | LST | LFT |
|---|---|---|---|---|---|---|---|
| 1. Repair rear wall | 20 | 42 | Share 6 | 38 | 58 | 80 | 100 |
| 2. Water wash | 13 | 0 | | 25 | 38 | 25 | 38 |
| 3. Restart Process | 8 | 0 | | 149 | 157 | 149 | 157 |
| 4. Cool down | 23 | 0 | | 1 | 24 | 1 | 24 |
| 5. Replace bad tubes | 54 | 0 | | 38 | 92 | 38 | 92 |
| 6. Close up firebox | 1 | 42 | Share 1 | 58 | 59 | 100 | 101 |
| 7. Close up water side | 9 | 0 | | 92 | 101 | 92 | 101 |
| 8. R & R relief valves | 50 | 27 | | 24 | 74 | 51 | 101 |
| 9. Fire-up boiler | 48 | 0 | | 101 | 149 | 101 | 149 |
| 10. Open Firebox | 1 | 0 | | 24 | 25 | 24 | 25 |
| 11. Shut-down | 1 | 0 | | 0 | 1 | 0 | 1 |
| 12. Clean good tubes | 23 | 31 | | 38 | 61 | 69 | 92 |

The above chart shows that certain paths through the project have float and others do not (and are therefore on the Critical Path). If several different paths do not have float, then all of them are on the Critical Path. If a path is delayed and the float is used then it joins the other paths on the critical path.

**Critical Path = A11-A4-A10-A2-A5-A7-A9-A3 which total 157 hours** All the EFT – LFT = 0 for any activity on the critical path. If any of the 8 activities on this path get delayed, the project will be delayed by a similar amount. The only way to avoid project delay is to intervene with one of the critical path compression strategies discussed below.

**Path A4-A8-A9 = 27 hours of Total Float.** It is calculated that by subtracting the EFT from the LFT for A-8, A-8 can be completed at any time up to hour 101 without having an impact on the project of any successor Activity. In this path, the float is total, free, and independent (meaning the activity does not impact either successors or predecessors until it runs out of float).

**Path A2-A1-A6-A9 = 42 Hours of Total Float.** In this example, A-1 has total float but no free float or independent float. If A-1 is delayed it will have an impact on A-6. If A-6 needed a special crew, that crew would have to await the completion of A-1 before it could do its work. If A-1 ran 16 hours over, it would not have an impact on the whole project but it would have an impact on the special crew for A-6.

**There are a series of related concepts to shorten the duration of a shutdown called crashing and fast tracking. Both these concepts are special applications of critical path compression.**

*Crashing:* Running the project with the shortest duration. The diagram shows that as the length of the shutdown is shortened, the costs are increased. There is a limit called the crash time. Any additional resource will not result in shortening the shutdown but will only result in wasted money. In fact, additional resources might slow the project!

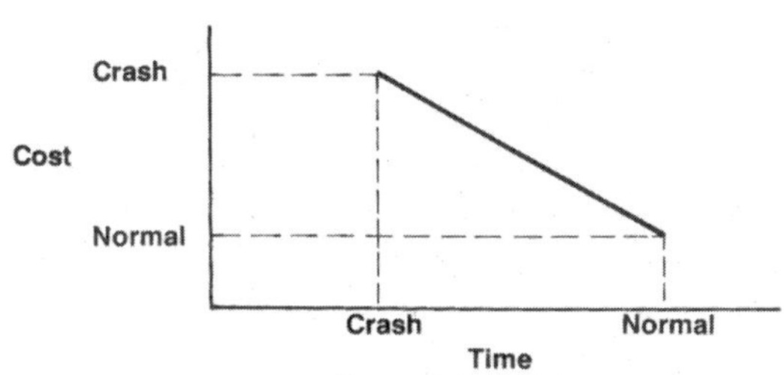


The process involves use of techniques to shorten the duration (irrespective of how those techniques impact shutdown cost). All the techniques of critical path compression are used as well as just adding resources (even if they will not be completely utilized).

In shutdowns, the cost of missed production may dwarf the cost of the extra resources to get the shutdown completed earlier. In the diagram, as you reduce the time of the project you increase the cost to perform that activity. The diagram shows a linear relationship, but actually, as the shortened crash time is approached, the costs increase at an increasing rate. At some point, increasing the cost will not shorten the duration of the project.

Related concept is **Fast Tracking,** which consists of doing activities in parallel that would normally be done in series. Fast tracking a painting job might mean starting the second coat of paint on the parts of the tower where the previous coat is dry, but before completing the first coat (running two, three, or more painting crews).

The most famous example of Fast Tracking was on the Manhattan Project in the 1940's. The construction of the Hanford Nuclear Facility began before all the engineering was completed. In fact, construction was started before the research into which process would be used was completed. Cement was being poured as the drawings for the very next section were being completed. The mission was to finish the project as quickly as possible without regard for cost. The thought was that while this was wasteful (entire factories were built and then abandoned when it turned out that the technology wasn't needed), the consequence of being second was unacceptable.

On an oil-drilling platform, critical paths have to be compressed for economic and practical reasons. The shutdown crews are stuck on board the rig for up to 3 months at a time (living off an accommodation barge). Unlike the permanent rig crew, these workers may not be resigned to living in those conditions. Things can get tense when there are 150 workers working in a space that is designed for 70 or 80.

To compress the critical path, heavy items are placed as close as possible to where they will go. The trick is to avoid the second lift and move the item without a big crane. Since the logistics for a lift are so complicated, pre-placing items might speed up the process by as much as a day or two. As much piping as possible is also prefabricated and placed right next to the areas where it will be used. It takes months of work (in addition to the time needed for planning) on a rig to prepare for a month of shutdown. But the time spent shortens the shutdown itself.

**Critical Path compression (or what to do when the critical path falls behind)**

- Reallocate resources from activities with the most float.
- Break activities up so that sub-sections can be done in parallel (fast tracking)
- Overlap tasks by creating partial dependencies (start activities before the predecessor is fully complete)
- Have crews pre-assemble or prefabricate the next items and bolt them into place
- Try to re-sequence to optimize use of facilities and time
- Increase overtime, add additional shifts
- Subcontract or increase contractor workforce
- Remove obstacles, make sure everything is there and ready to go, check the next few activities and make sure nothing will hold them up
- Increase facilities; increase the number of cranes, size of equipment.
- Move activities off the critical path to next shutdown (if that is possible)
- Make the job smaller (reduce scope of any activity on the critical path)
- Use incentives such as extra money for early completion, extra days off, create games and competition between teams,

# Chapter 16

# Resources

One of the breakthroughs made by modern project management techniques is the ready, instant, and ongoing analysis of resources. One of the major causes of shutdowns running overtime is inadequate resource. Now, using modern PMS, an individualized calendar can be generated for any resource (that you listed when you set up the activity). Most resource conflicts can be seen and resolved at a glance.

PMS systems treat resources similarly. It is kind of strange to consider pipe fitters, torque wrenches, and yards of concrete as the same kind of commodity, but they are all resources that can be consumed during a shutdown. All can be purchased for a shutdown and have unique constraints. The pipe fitter, John, can only work 12 hours a day, 7 days a week (for a while), the torque wrench can be used 24 hours a day and 7 days a week. The concrete has to be used within an hour of being mixed.

There are two classes of resources, which are very different. One class is resources measured by hours such as labor, crane usage, or tool usage. The other class is resources measured by weight, quantity or volume. This would include spares, concrete, pipe, etc.

There are two major attributes of resource usage, quantity, and sequence (or timing). The challenge (for resources consumed by weight, length, etc) is to have enough at the beginning of the shutdown or delivered during the shutdown, but not too much. When the resource is perishable, or takes up space, the timing becomes more important. When you order mixed concrete you always specify how much, and especially when you want it. No amount of shifting activities will change the need for 4-inch pipe. If you need 1100 feet of 4-inch pipe, and you have the space to store it, you need that much (preferably when you start the shutdown); one foot less, or one day late, and the shutdown will be delayed. Speeding up the activities will generally have no impact on the amount of pipe needed.

If the resource is measured in hours, it is important to have enough for the peak need and to be able to shed the resource (so it doesn't have to be paid pay for), as soon as you don't need it anymore. If the resource is measured in hours, then the peak quantity can be manipulated by shifting activities within their float. The important attribute here is sequence. This rule answers the question of when do I need the resource? The rule has particular application for heavy equipment, people, and things that are scarce or expensive (such as infrared cameras). Shifting resources may change the peak need for the resource (but will not change the total hours for which the resource is needed).

| Resource | Comments |
|---|---|
| Labor hours by skill set or craft | Labor costs are the first thing people think about when discussing resources. Be careful with more exotic sub-specialties, and break some crafts down further to skill sets. Shift activities so that the number of people needed is as constant as possible. |
| Parts | Large expensive parts with long lead times are usually on everyone's radar. The shutdown timing might be dictated by the delivery of these parts. Be sure there is a place to store the items that will protect them. Small parts are important too. Shutdowns have been shutdown by lack of bearings, fittings, and other minor items (particularly if a similar plant shutdown occurred just before yours and wiped the distributors in the area clean). |
| Materials and supplies | Similar considerations apply as to small parts but the lack of welding wire, rags, tape, wire nuts, etc can be a high cost inconvenience. Usually someone can jump in a truck and pick this kind of stuff up, even if it has been wiped out of the immediate area (of course this is a big issue if you are located in the middle of the ocean or not near a supplier). |
| Large equipment | As with large parts, heavy equipment is usually intensively managed. The constraint is when two crews need the crane at the same time, which should be handled by the schedule. This item also is measured by the hour, so playing with activity timing might solve the access problem. The second issue of this resource is to return it the day you no longer need it (be sure to make clear arrangements because getting an item back might be pricey and embarrassing). |
| Tools | Sometimes common tools can be a bottleneck. A hundred torque wrenches that are needed for a few days might present a problem for some organizations. Be careful, because many maintenance plans do not go to this level of detail. The excessive need for such a resource might be hidden until the day it is needed. Cost of such a resource is by the hour, so shifting activities within their float might reduce the peak number of wrenches needed. |
| PPE | Resist the temptation to go ahead and do the job without protection if there is not enough PPE (personal protective equipment) to go around. This choice can be deadly so be sure whatever PPE needed is called out on every job. PPE is used by the hour so that shifting the activities might help. |

The Resource conversation must go on with both the logistics and the human resource side of the shutdown. If any resource element of the planned job does not converge at the right place at the right time, the job and potentially the whole shutdown will be put into jeopardy. The point people, the people charged with responsibility of managing a section of the shutdown or an activity, are primarily tasked with keeping on top of all types of resource requirements and intervening when they anticipate an upcoming problem.

## Crude Hourly Resource Analysis

Early in the shutdown development (Phase One) someone might have a need for rough numbers for budget purposes. Knowing what hours will be contracted and what will be available from existing crews is very useful.

| Resource | Shutdown Hours Required | Available from your crews | Hours needed by outsiders |
|---|---|---|---|
| Electrician | 4250 | 1050 | 3200 |
| Instrument Tech | 1600 | 1800 | 0 |
| Pipe fitter | 12,500 | 3600 | 8900 |
| Mechanic | 10,800 | 6500 | 4300 |
| Welder | 1200 | 600 | 600 |
| Labor | 32,000 | 6000 | 26,000 |

This table does not take into account when the resource will be needed but does point to a general direction and scale of the contracting effort.

## Resource Leveling

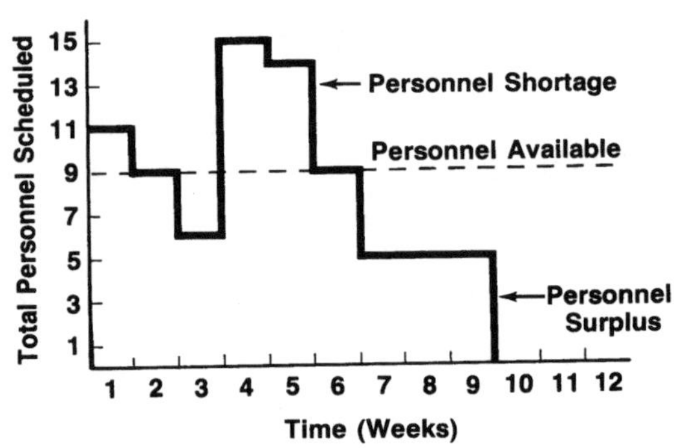


**Image from Defense Systems Management College, 1990 Scheduling Guide**

If the drawing above showed the resource requirements for a shutdown there would be a problem. It would be necessary either to hire and fire on a weekly basis, over-hire and hope other work can be found, or under hire and delay the completion. The other alternative is to level the resource. Keep in mind leveling the resource can apply to people, fork trucks, cranes, welders, and any other hour-limited resource (but not materials that get consumed by pieces, weight, volume, etc).

One of the powers of modern PM software is to aid in shifting resources within their float to level the need.

Resources are needed irregularly throughout the duration of the project. At some times the resource requirements may exceed the amount of resource available. Conversely, the resource might not be needed at all during certain times in the project. We have to keep the outcome (project completion), lowest cost, and the critical path in mind, whenever we look at modifications to resource usage.

In the example below, 12 pipe fitters are needed on day three and none are needed on days five and six, Presenting a difficult but common scheduling situation. The dark lines show critical activities and the dashed lines show float. The next three figures are courtesy of Mike Brown (at NSI in resource section).

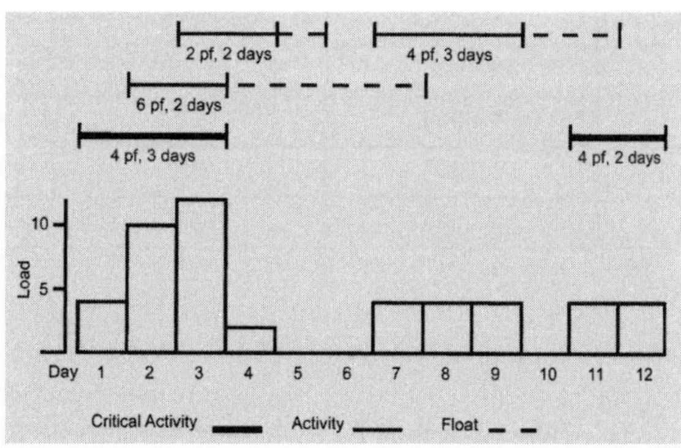


As you can see from the figure below, the first attempt at shifting the jobs within their floats results in a significant decrease in variability of the need for Pipe fitters.

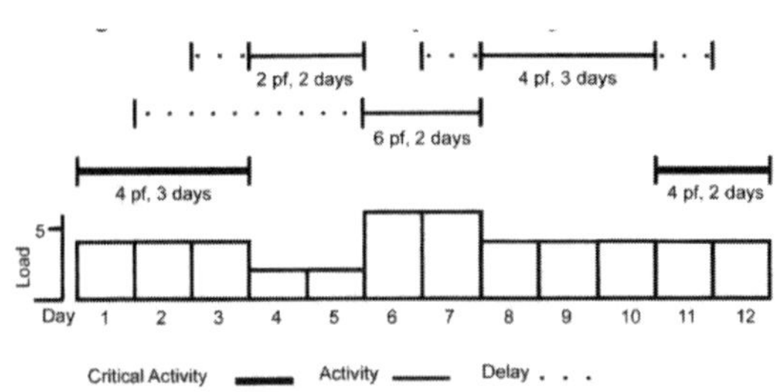

This final leveling shows more advanced techniques. In addition of moving within float one job was slowed down by assigning fewer pipe fitters.

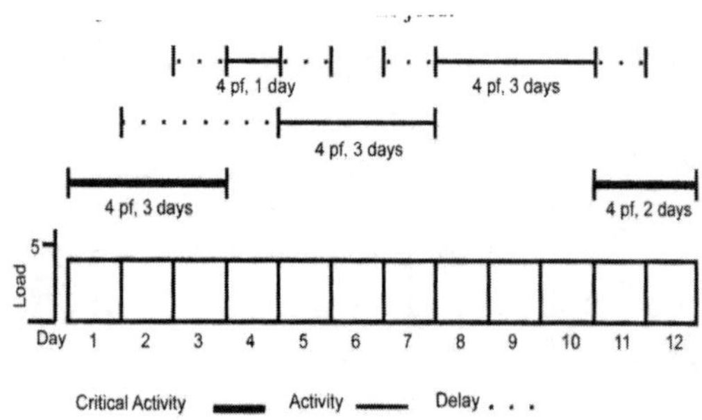


This last manipulation produces the ideal scenario (constant demand for a labor resource). An hour spent moving jobs and their resources could save hundreds or thousands of dollars.

**An approach to resource leveling**

1. Start out with the assumption that all activities start on their EST
2. Day 1- create a list of all activities that are on a critical path until you either run out of activities or run out of resources. The next step would be to assign specific resources and times to each job making your list into a true schedule.
3. If you don't have the resources for the critical activities, look at increasing resources (you will have to do something like adding overtime, adding people, giving helpers to skilled workers – whatever works.)
4. If resources are available, complete listing of all critical jobs, then list non-critical activities. Start with activities having the smallest float, and continue until you run out of activities or resources.
5. If you run out of activities and still have resources you have a surplus of that kind of resource for that day. Look for jobs you can accelerate in some way to absorb these resources, or use them as helpers for other trades.
6. List the critical activities for the next day. If resources are freed up by completion of any job, return them to the pool.
7. Follow through with 2, 3, 4 and 5 for day two.
8. Continue until critical activities are completed.

To preserve Resources or level resources

- Move the task within its available float so that the same resource (for example: a person or a crane) can do both jobs.
- Extend the duration of the activity within its available float by reducing the resource input (for example using 4 pipe fitters instead of 6)
- Interrupt non-critical jobs and restart them when resources become available (only if necessary, since productivity is lost and you open the possibility of either forgetting the interrupted job or allowing contamination to ruin the interrupted job).
- Reassign work to resources not otherwise occupied
- Front load or back-load the activity when the resource is free (to reduce the resource requirement during any critically overscheduled period.
- Add new resources (are there any sources of resource not considered before –for example can you borrow resources for short periods from other plants?).
- Expand the workday or increase critical resources from 8 to 10 or from 10 to 12 hours a day (as mentioned there is a cost to this in productivity, quality, and safety.)
- Schedule overtime
- Use a support resource. If you have underutilized human resources, use them as assistants or go-fers to expedite the critical resource.
- Reduce scope or drop non-critical jobs that use the same resource
- Accept new project completion date.
- Yell more (just kidding)

# CHAPTER 17

# PERT AND NEWER METHODS

PERT requires three estimates for the duration of each activity: the optimistic, most likely, and most pessimistic. The three estimates are factored together to determine the most likely project path.

$$t_e = \frac{t_o + 4t_m + t_p}{6}$$

Where $t_e$ is the Expected time,
$t_o$ is the Optimistic time,
$t_m$ is the most probable activity time
$t_p$ is the Pessimistic time.

The most probable time is weighted (4X). The weights are based on an approximation of the Beta distribution.

The three factors that describe the estimate are the $T_e$ or the expected time of completion, the standard deviation that describes the variability of the estimate. Variability can be described as the thinness or fatness of the beta curve in the next figure. A high standard deviation would produce a fatter, wider, curve and a low standard deviation would describe a narrow, thinner, curve. The variance is the third factor, which helps us calculate the variability of a whole path through the project.

For example: we are executing a simple Boiler Shutdown and the contractor says that Activity 12 (Clean good Tubes) will probably be done in 23 hours. She adds, if everything goes right it will be complete in 18 hours and the worst case is 36 hours. According to the formula the expected time is: 24.3 = (18 + (4*23) + 36) / 6.

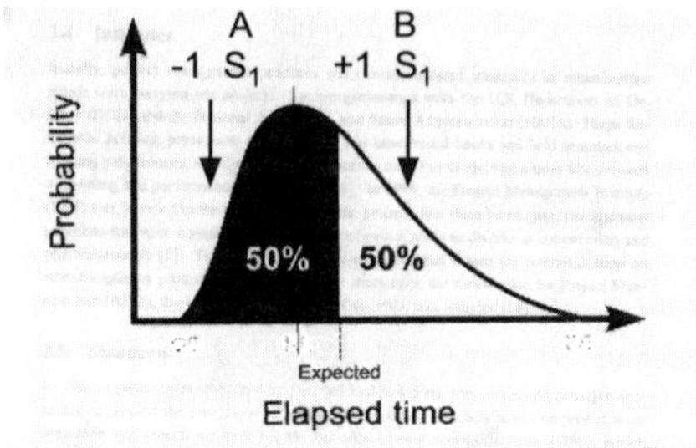


## Calculating the Standard Deviation

The Standard Deviation (SD) or $S_1$, is a measure of the variability of the estimate of each activity, and is closely approximated by the simplified formula:

$$S_1 = \frac{t_p - t_o}{6}$$

In our tube cleaning example the SD is 3 = (36 – 18)/6

The Variance is the Square of the Standard Deviation or VAR = $9 = 3^2$

## The power of statistics

With these calculations we can use the power of statistics to calculate the project duration and the probability of the project being done in the time available. One of the assumptions of PERT is that activities are independent, and that the time required to complete one activity has no bearing on the completion times of its successor activities in the network. This assumption is considered to be one of the weaknesses of PERT because problems with completing one activity can certainly be related to problems completing another. For example, if an electrical activity runs behind schedule, and the same electrical contractor is responsible for other activities, we may find those other activities are delayed.

PERT also assumes that the expected length of a project (or any sequence of independent activities) is simply the sum of their separate expected lengths. The sum of all the $t_e$'s along the critical path gives the length of the project.

Similarly the variance of a sum of independent activity times is equal to the sum of their individual variances. To see how variable the estimates are on a path through the project, simply do the calculation for the variance for each activity and add them up. In our boiler example, the sum of the variances of all the activities along the critical path A11-A4-A10-A2-A5-A7-A9-A3 is found to be 100 hours.

**The square root of the variance gives the standard deviation.**

$$\sqrt{\text{VARIANCE}}, \quad 10 = \sqrt{100}$$

The higher the standard deviation, the greater the variability. The greater variability leads to increased un-certainty that the project will be completed on the due date.

Although the $t_e$'s are randomly distributed, the average or expected project length $t_e$ approximately follows a Normal Distribution. We have a lot of information about a Normal Distribution, so we can derive several statistically sound conclusions from these calculations. These conclusions help the project manager report not only the expected completion date but also his or her confidence that it will be completed by then.

## How to calculate the probability of the shutdown duration

A random variable drawn from a Normal Distribution has 0.68 probability of falling within one standard deviation of the distribution average (in our example between A and B on the beta diagram). Therefore, there is a 68% chance that the actual project duration will be within one standard deviation, ST of the estimated average length of the project, $t_e$.

In our example, the $t_e$ = 157 hours and the ST = 10 hours. Assuming $t_e$ to be normally distributed, we can state that there is a probability of 0.68 that the project will be completed within 157 ± 10 hours, which is to say, between 147 and 167 hours. If we are NOT concerned with running shorter than predicted, the probability that the shutdown will be 167 hours or less is about 84.9% (the sum of the probabilities from B on the beta diagram to the Y axis or zero)

It is known that just over 95% (0.954) of the area under a Normal Distribution falls within two standard deviations, so the probability that the project will be completed within 157 ± 20 (or 137 to 177 hours) is very high at 95%. Given these estimates and information, the project has a 97.7% chance of being completed in less than 177 hours.

**Of course all these calculations are only as good as the estimates!**

One of the traps of the PERT method is that project managers become enamored with the seeming accuracy of the calculations. They lose track of the fact that if the estimate is $\pm 15\%$ then the accuracy of the completion time will be closer to 15% than 0.5% (Partially adapted from an article by Ajit Mani founder of Intervention PVT. LTD)

## Other Functions

The three-estimate method (and weighting the expected by 4) produces a result called the Beta distribution. Project managers have also used other approaches including the normal, Binomial, Poisson, and Weibull distributions. If the organization has significant experience in the particular repair, and has good records, a special function built from actual data can be used.

Other approaches such as the Monte Carlo method, that might yield more accuracy in extremely critical projects, are also used.

### Variability is well known in some kinds of activities on shutdowns

Sometimes standards can be set for the tolerance band for different types of jobs.

For example, in construction, Reddish recommends in *Managing Projects:*

| | Most optimistic subtract | Most pessimistic add |
|---|---|---|
| Excavation | -10% | +20% |
| Fabrication and erection | -5% | +10% |
| HVAC | -5% | +5% |
| Mechanical | -5% | +15% |
| Electrical | -5% | +5% |
| Revisions | -10% | +20% |

## GERT (Graphical Evaluation and Review Technique)

A modification of the PERT is GERT. There were problems with the PERT technique when it was put into unusual project management circumstances. GERT techniques are similar to PERT but have the distinct advantages of allowing for looping, branching, and multiple project end results.

With PERT one cannot easily show that if a test fails it may have to be repeated several times unless it is known exactly how many times it is planned to repeat the test.

With PERT, it cannot be shown that, based on the results of a test, one of several different branches can be selected to continue the project. These problems are easily overcome using GERT. (Adapted from PM Forum library glossary. URL is in resource section)

## Critical Chain

*"Work expands to fill (and often exceed) the time allowed." -- Parkinson's Law*

*"Whatever can go wrong, will." -- Murphy's Law*

These two rules are well known to maintenance shutdown professionals. Eliyahu M. Goldratt designed a modification of the PERT system based on his theory of constraints that takes these two principles into account.

Critical Chain is an alternative to critical path that when used correctly will give more accurate representations of project duration, because it takes into account the natural tendency toward the issues described in Parkinson's and Murphy's laws.

Every estimator adds time and resource (called buffers) into estimates to protect themselves from small problems. Critical Chain looks closely at these buffers. The identification and management of all these buffers is the core of critical chain theory. In simplistic terms, the practice removes the buffers from individual activities and adds them to the end of the path.

This procedure avoids wasting the buffered time due to Parkinson's Law, which says that work will expand to fill the time available. Any time left over in an individual activity will be squandered. The time will be consumed as the activity expands to fill the time allotted (that's why the short estimate in PERT is almost always exceeded).

In Critical Chain project management, the buffers are removed from individual activities and added to the end of that chain. This is where Murphy's law (if anything can go wrong, it will) comes in there will be time at the end of each path for the problems presented by Murphy's Law to be handled.

Resources normally are plotted separately in bar charts or other documents, but in Critical Chain the resources are the drivers, and less effort is put into optimizing the network. In Critical Chain a "good enough" solution is adequate because the inherent uncertainity in estimates is much greater than the difference between the optimum and near-optimum network.

# CHAPTER 18

# PM (PROJECT MANAGEMENT) SOFTWARE

Once you get used to using software to lay out shutdowns it is unlikely you will go back to manual graph paper methods. Simple shutdowns can be plotted in a day or less, and you have the advantage of nice, easy graphing.

Most commercially available PMS were designed for general-purpose project management. Users primarily used the software to manage software projects, building construction, and other standard type projects. These standard systems may have to be configured to work for shutdowns. For example, MS Project allows the addition of fields so that work order number and asset number can be added to the task. This facility simplifies relating PMS activities to CMMS work orders.

From the history of PERT and CPM it is apparent that today's project software looks the way it does as a result of the problems those programs were trying to solve. PMS was originally designed to compute the Critical Path through a project and later to calculate resources and graph all the relationships implicit in the project. As a result, modern PMS has significant capabilities to:

1. Calculate the critical path and float
2. Create Gantt charts and network diagrams
3. Communicate shutdown parameters to management and contractors
4. Communicate expectations to contractors and supervisors
5. Calculate resources based on inputs
6. Provide aids to balance resources
7. Track and display milestones
8. Do quick and dirty estimates and track changes when assumptions change

9. Calculate 'what ifs' when assumptions or constraints are changed
10. Provide visual displays of different shutdown parameters
11. Print a series of canned reports like baseline to actual, resource allocation, cost distribution, etc.

There have been some attempts, with varying levels of success, to address the shortcomings of the present systems, but if the PMS has capability in these areas it is still pretty basic. General purpose PM Software is not as useful for:

1. Job Specifications (some specialized vendors provide databases with all kinds of specifications for generic work in refineries or chemical plants to help here)
2. Scope of work issues (again some specialized vendors provide databases with pre-written work scopes to help here)
3. Managing performance beyond reporting what has happened
4. Providing an easy way to manage feedback about job completion and partial job completion.
5. Quality control
6. Keeping the manager's eye on the ball (instead, it may fill his/her mind with paperwork and charts, and ignore reality.
7. Managing Contracts

One PMS vendor that has specialized in Shutdowns that add significant capability and expertise is Interplan Systems. Information about that firm can be found in the resource section.

**Some Functions to look for in your PM Software**

- Ability to maintain a list of work orders with direct or indirect links to the work management system or CMMS.
- Capacity to develop both network diagrams to show logic (CPM) and Gantt charts to show duration
- Comparison of time relationships to deadlines (how much or how soon until milestone)
- Calculation of the critical path
- Ability to compare current status to planned status at that date
- Provision of cost projection reports
- Indication of float (total, independent, etc)
- Indication of milestone status
- Provision of milestone reports
- Assistance in resource planning, resource leveling
- Production of summary reports for management

- Assistance in computing alternative courses of action

**Some other functions to look for include:**

*File import and export:* It is important to get the work list information from your CMMS and send it to your PMS.

*Macro capability:* Useful to automate common functions (similar to macros in Excel).

*OLE (Object Link and Embed):* This is a fancy Windows® term that lets data be cut and pasted from one program to another. One of the most useful Cut and Paste capabilities is between Excel® spread sheets and MS Project®. Just about all CMMS can export shutdown work lists to Excel®. Project® lets you Cut and Paste from Excel®. That one capability could save hours of typing.

**Before choosing software, consider your organization.**

Answer some questions before software is even examined. Will the use be intensive or casual? Is there a dedicated core of shutdown people who can become experts in the package or is shutdown a casual annual job? These questions should be answered before starting to search for a system so that answers will be ready for basic questions that vendors will ask, such as the following.

- How often do you have shutdowns?
- List the shutdown functional managers (such as Logistics, Engineering, etc.) and their teams and describe their proposed interaction with the PMS.
- How far apart are these people's offices: same office suite, same building, same campus, same city, and same country?
- Would contractors need access to this system for viewing, for data entry or for making changes?
- How many people would need viewing access to the PMS (just be able to look at the charts)?
- How many people would need access to change or enter data in the PMS?
- Does management (policy committee and other top managers) require sophisticated charts to oversee the shutdown progress?
- Do you have clearly different shutdowns like a 2-day semi-annual inspection shutdown and a once every 4 years, 30-day, major overhaul?
- What is the average duration of the last few shutdowns (by type)?

- What would you estimate is the average spent for labor and parts on the last few shutdowns (again keep the types of shutdowns separate)?
- How much money is taken up by shutdowns?
- What is the scope of the shutdown, outage, and turnaround operation?
- Do you have a partial or complete shutdown of your facility or both?
- How often do these events happen?
- Do you have people permanently assigned to shutdowns?
- Do you have a CMMS? What type? Does it have links to any PMS already built?
- Did you use contractors to manage shutdowns? If so, it is likely they will already have shutdown software.
- Do you currently use software to help plan and organize the shutdown? Is this search an upgrade?

Not every firm needs all the capabilities of the various systems. In PMS, as in Word processing, vendors load up their software with all the capabilities they can to make it appeal to the widest possible market. Few firms use all the capabilities. But there are essential capabilities for your shutdown situation and your organization.

## Selecting PM Software

There are 50 or more packages available for project management and a few specifically for shutdown management. A simple search of the web for "Project Management Software" will yield a good number of leads. There are also sponsored links for PMS software companies.

Cost of these of packages run from US$50 to $500,000 so there are a wide variety of levels, capabilities and support levels. Several vendors are listed in the resource section but they were chosen to demonstrate various levels of system rather than because they offer the best systems. Most of the vendors offer free demo software from their web sites.

The most popular is Microsoft Project, but there are many others, some of which are designed specifically for shutdown type projects.

Primavera (one of the primary vendors of the larger PMS) has built links between the Maximo CMMS package (where the shutdown work orders are stored) and their PMS package. When a work order in Maximo is closed out, Primavera completes the activity and updates the schedule and critical path.

Most CMMS can export a list of shutdown work orders to Excel. Microsoft Project (as well as others with OLE) can accept a selected part of the spreadsheet using the paste function to build an initial project plan. One key is to add a field to Project for the work order number so you can track the action.

**Chart to help identify some of the important issues and uses of a specific package.**

5= Very Important, 4=Important, 3=somewhat Important, 2= unimportant, 1= N/A

| # | Item | 5 | 4 | 3 | 2 | 1 |
|---|---|---|---|---|---|---|
| 1 | Need user friendly for non-professionals to interact and enter data | | | | | |
| 2 | Need CPM capability (Critical path and float) | | | | | |
| 3 | Need PERT (3 estimates on duration) | | | | | |
| 4 | Need sophisticated duration capability such as Monte Carlo, other functions, actual data | | | | | |
| 5 | Need to identify resource conflicts and resource lulls | | | | | |
| 6 | Need rapid development for possible projects | | | | | |
| 7 | Need big picture graphical views for presentations | | | | | |
| 8 | Need milestone charts for management | | | | | |
| 9 | Need alerts built in to notify PM operator of potential problems | | | | | |
| 10 | Need to be able to quickly rerun charts to manage risk events that throw off the critical path | | | | | |
| 11 | Need network capability with appropriate levels of security so that many people can see the data but only a few can modify it. | | | | | |
| 12 | Need to sort out activities and paths to be printed for supervisors and shutdown teams in action formats | | | | | |
| 13 | Need to be able to send sub-sets of the data to stakeholders easily | | | | | |
| 14 | Need budget verses actual costs in near real time | | | | | |
| 15 | Need interface to ______ CMMS to set up shutdowns | | | | | |
| 16 | Need interface to ______ CMMS for activity (Work Order) closeout and synchronization | | | | | |
| 17 | Need interface to _______ ERP system for accounting information such as incoming invoices, payroll, etc. | | | | | |

**When looking at the vendors (particularly on the larger and more expensive systems) be sure to ask if:**

1. Does the system run on standard computer hardware (not special hardware incompatible with everything else). Is the system compatible with existing LAN?
2. The better choice is a PMS that can run on the company Intranet and has web portal capability. Ask if it can handle TCP/IP communications protocol.
3. If you are buying a major system, find out whether the vendor has the financial strength to complete the contract (and stay in business for several years after installation).
4. Find out if the vendor has software support people, and check whether you can get through to a live person easily. Is there a toll free phone number? Check whether the people know the product and something about shutdowns?
5. On large systems, find out if the vendor can provide economical customization?
6. Ask if the vendor has an on-going enhancement program, and how often versions have come out in the past.
7. Investigate whether the programmers are employees of the vendor or contract workers?
8. Ask whether the vendor has a local installation organization or a local VAR that can install and show you how to get the most benefit out of the system?
9. Check on the vendor's experience in management of shut-down projects of the size of yours?
10. Find out if the vendor's technical people are well cross-trained (in software, hardware, and reality ware, such as how a shutdown is different from a construction project)?
11. Determine that the vendor has been in business 5 years or more?
12. Is there an active user group independent of the vender or news group for the package?

# CHAPTER 19

# MASTER SCHEDULE FOR PHASE 2 PLANNING

### Design and Engineering Phase
### Phase 2 Comment Box

### Scope

Phase two begins with management of the scope of the shutdown. In Phase 1 the top management has outlined the broad-brush scope of the project. In Phase two the scope is refined to be actionable. In other words the scope is broken into sub-projects called activities that can be managed, overseen, contracted out, and planned individually. Planning is completed in phase 2 on two levels. Each individual job is planned and all the planned jobs are assembled into a shutdown master plan.

### Structure for the management of the shutdown

The overwhelming bulk of the work of project planning is in this phase. The goal here is to create a detailed structure to manage the shutdown. By the end of this phase resources start to be put in place. The design and engineering phase includes drawings, planning documentations, and firm budgets. All stakeholders have had input. Planning, drawing packages are available for review and bids

### Manage Risk

Detailed risk analysis is completed in this phase. Contingency plans are designed and put in place before execution. Whenever major changes are made to any part of the plan, the risk analysis should be rerun. Management of risk is one of the reasons why so much trouble is devoted to planning of outages. When risk is identified ahead of time there is an opportunity to manage, reduce, eliminate, or accept it. A Job Safety Analysis (JSA) is included in the individual job plans. The method insures that safety is built into the shutdown and not added on at execution time. At the same time, planning is done to insure quality.

| Milestones | |
|---|---|
| Project Work Breakdown Structure (WBS) | This item alone could take weeks of work |
| Project Management Software (PMS) | Project management software (PMS) can produce several kinds of charts including Gantt, PERT or CPM from the same breakdown structure. |
| Project Cost Estimate | In Phase I the estimate is made for approval and in this phase the final estimate is calculated. The PMS can tally costs as the projects are entered and details become known. |
| Resource Requirements vs. Resource | It is critical to think through the number of people needed to deliver the project on time. PMS can calculate personnel requirements (as well as the need for other potentially limited resources). |
| Team is in place | By this time a physical structure (schedule with all work orders) has been put in place, plus a personnel chain of command to manage the project. Experts in each area, contractors, engineers and supervisors have been chosen for this shutdown, and they have now reviewed this version of the turnaround schedule. |
| Planning is completed Plan as in the whole project and plan as in individual jobs | Large jobs have been broken into activities. Details are being added to fill in the dark areas on the schedule. These tasks, resources, and dependencies should be entered into Project Management software. The plan can be printed as a Gantt chart and managed as a network diagram |

Your question as you read through this list is "Do I need this item on my master list for Phase 2?" Your starting point is a subset of this list that completely reflects your shutdown enviroment.

| Do I need this? | Action Item |
|---|---|
| | 1. The policy team approved the shutdown. Now start the process to adjust the scope of work to agree with the budget approved |
| | 2. Create a Project Timeline that starts now (with the planning phase) and ends at the conclusion of the shutdown. Choose milestones for completion of engineering, planning, procurement, contracting, work scope, safety analysis, and logistics. In short, establish a milestone for everything that is essential to the success of the shutdown. Some firms call this a level 2 schedule. Enter milestones into PMS (Project Management System). |
| | 3. Authorize engineering for all modifications or projects. A large project will need intermediate milestones such as engineering drawings are 30% done, 70% done and 100% complete and approved for production. |

| | | |
|---|---|---|
| | 4. Perform or schedule PM and PdM inspections to determine the extent of work and the possibility of extra work that might be needed. Accelerate PMs that are "almost" due. PM is targeted to find deterioration. Deterioration can impact the shutdown scope. | |
| | 5. Begin process of validation of the work list. This list will be revisited many times before the shutdown begins. | |
| | 6. As the general scope of the turnaround stabilizes, draw up a preliminary high-level schedule for use at the first meeting and to determine workforce requirement, number and type of supervision, and proportion of contractors. | |
| | 7. Begin site plan design. As the scope is stabilized, plot out major storage issues including staging areas, set down and rebuilding areas, crane storage, crane placement for major lifts, mobile equipment parking, storage trailers, office trailers, assembly points, etc. If you have severely limited space the site plan might be more of a series of plans, changing in time. | |
| | 8. Begin data entry of jobs to PMS. Where possible at this early stage, break down large jobs to activities and into WBS. | |
| | 9. Pre-qualify contractors who will be bidding the job. | |
| | 10. Define what services you will supply to the contractor including water, compressed air, electricity, waste handling, security, lighting, and where/how they will be supplied. | |
| | 11. Decide whether you or the contractor will be responsible for logistical support of their employees. Push as much responsibility on to the contractor unless you can supply it much cheaper. This is spelled out in the contract and discussed in the pre-bid meetings. | |
| | 12. Based on the proposed scope and schedule write a scope of work and bid packages that covers the major points but is flexible enough to not paint you into a corner. There might be fixed price bids for large tasks, T & M bids for small tasks, and unit bids for bulk works. | |
| | 13. Take photographs of the site elements that could help a contractor visualize the major parts of the job accurately. These 'before' shots make a great addition to the shutdown narrative report. | |
| | 14. Bid the job, review responses, and choose a contractor. Write a rule into all contracts that all invoices have to be tendered within 4 weeks of the end of the shutdown unless prior approval is obtained. | |
| | 15. Give the contractor preliminary indication of the start date, duration, number of people, and skills needed. This data will have to be reviewed as final issues are pinned down. | |
| | 16. Order long lead-time spares. | |
| | 17. Place orders for long lead time rental equipment. | |

| | |
|---|---|
| | 18. Turnaround office: Request temporary office space with all the trimmings for your staff if necessary. Help the contractor handle this for him or herself.<br>Include: Copying machine;<br>A computer(s),<br>Printers,<br>Network wiring, connection, and passwords<br>Internet access<br>E-mail accounts<br>Codes for access to CMMS<br>Provide software for project management<br>Software for writing reports, spreadsheets and paper forms, and office supplies for turnaround office<br>Desks Chairs and tables for coordinator, materials, and zone supervisors.<br>Additional phones.<br>One company project team moved in with the contractor for the planning and execution phase of the shutdown. In fact the major subcontractor moved in too. The company felt this was the best way to partner with the contractor to insure nothing fell through the cracks. |
| | 19. Review any long lead-time prefabs. Place orders if appropriate, otherwise set a date into the schedule for all prefab orders to be placed. Decide whether to make (in-house) or buy (outside vendor) each job. |
| | 20. Review bulk works (such as mass rebuilds of relief valves). Have initial conversations with vendor about their capabilities and timing. See if there will be any conflicts with resources such as rebuilder schedules, or removal and replacement labor. Place orders. Evaluate stock to see if you want to increase levels. |
| | 21. If you are responsible for them (try not to be) order scaffolding, portable toilets, nurse's stations, eye wash stations, showers, trailers, barricades, and other bulk items early, if required in your location. |
| | 22. Conduct job hazard safety analysis (JSA or JHSA) if appropriate on the jobs that make up the shutdown. Review the entire safety plan based on JSAs completed. On some jobs this step is the responsibility of the contractor. |
| | 23. Request safety department prepare for mass basic safety orientations of contractor crews. Include estimates and proposed timing. On some jobs a training contractor is chosen for the mass training. |
| | 24. Define and source any job-specific safety training. |
| | 25. Inform all craft supervisors as appropriate concerning getting equipment ready for use. |
| | 26. Announce and hold initial shutdown meetings with stakeholders to inform them of which jobs made the cut and which didn't. Also discuss the preparation needed for the event and approximate timing and duration. |
| | 27. Shutdown work list is closed. No new work is accepted. Work that 'must' be done will now flow through the special approvals process (if necessary, be taken to the Policy committee). |
| | 28. Work validation (which has been on-going) is completed for now. All extra work has to be validated against approved work lists. |
| | 29. Validated work is categorized and packaged. |

| | |
|---|---|
| | 30. Decomposition of more complex jobs is completed. This is the process of breaking jobs into specific activities to suit the project management software. (A useful aid is a previous shutdown in the same area, which can be used as a template or starting point for the activity list. Particular attention should be paid to activity sequencing and to any problems that occurred in the last job.) Enter all activities into the PMS. Determine time and effort level (called resource requirement) for each activity based on the estimate. The effort level would include the human resources used (trades or more specifically skill sets), the actual number of people, the labor hours, and the elapsed calendar time |
| | 31. With the work list closed you can complete the site plan with locations of major rental units, major spares, marshalling areas for prefabs, work areas, assembly areas for fire, etc. This drawing is part of everyone's turnaround package. It could be kept up-to-date on the shutdown web site (we can wish can't we?) |
| | 32. Plan for, and rent logistics trailers or buildings for parts storage are based on the size and quantity of spaces (some may be moved right next to the equipment scheduled for work). If the contractor is responsible, make sure this work is done. |
| | 33. Work through each major project and act out on paper the movement of materials, people and equipment to see if the plan is safe and reasonably efficient. Remember that in large jobs, collisions in labor, materials, or equipment is 20-1000 or more times cheaper to resolve on paper than on the ground. Note what jobs are running around the target job look for conflicts in space or materials (flammables used near welding). Test for jobs that occur at the same place on a level or vertically. Insure that these jobs are not scheduled at the same time. Check circuits to see if de-energizing will adversely impact other work. On very high-risk work consider practicing (dry run) until you are convinced the job will go smoothly. |
| | 34. For each project (or work order), detailed and complete risk identification, risk quantification, and risk response would be completed at this stage. Identify hazards for each job. Investigate risks generated by interaction between jobs, jobs on the same system, and jobs near each other, etc. |
| | 35. Is the work a critical operation on a critical component? If it is, and the stakes are high enough, do what NASA does. Whenever they have a repair (like the ones on the Hubble space telescope) they go into their giant swimming pool (which simulated space conditions) and practice over and over. Go a step further than a try on paper. Set up an area where you can practice until you have the job down perfectly (knowing that in the real world something will always happen but practice always makes it easier to manage the problems). |
| | 36. Calculations need to be made to determine the resources required for the job. How many contractors are needed? How many in-plant employees will be seconded to work with contractors? A balance needs to be drawn between resources needed to meet schedule and resources available. Run project management reports and play with float to balance resources. |
| | 37. Are there any resources (people, large equipment, even smaller equipment where it makes sense) at sister plants that can be borrowed for this shutdown? Many organizations encourage these kinds of transfers because they cross-pollinate the crews, provide a change of scenery, and give potential overtime opportunities. Do unrelated plants in the neighborhood share any resources (if not, should they)? |
| | 38. Arrange for an extra dumpster(s) for waste. Arrange for liquid waste trailers if needed, if you are responsible (try to make the contractor responsible for as much as possible). Keep in mind that you are responsible for waste from 'cradle to grave.' See that material and tool trailers are properly supplied. |
| | 39. Review PPE requirements with contractor, keeping in mind that PPE can be inspected at any time and any contract employees that are not in compliance can be removed from the site. |

| | |
|---|---|
| | 40. See that room and transportation arrangements are made for employee supervisors when there is a need for a change in shift coverage. |
| | 41. Make lodging and transportation arrangements for foreign workers (such as specialists on off-shore equipment) and sub contractors coming from a distance |
| | 42. If the facility is large, consider leasing a shuttle service from major points (like the front gate to the contractor's trailer). |
| | 43. Generally on longer jobs the contractor can make transportation arrangements efficiently but on small, short lead-time jobs they might be your responsibility. If it is your responsibility, arrange for transportation, housing or personal needs for crews as applicable. If it is not your responsibility, verify that appropriate arrangements are being made. Ask questions about the following:<br>Temporary eating and food prep areas including vending machines<br>Temporary first aid<br>Temporary showers and changing/locker rooms (clean and dirty- Men's/Women's)<br>Temporary accommodations<br>Temporary storage<br>Smoking areas<br>Social hall, entertainment (karaoke room?)<br>Pay phones<br>Internet access |
| | 44. Arrangements for mobile equipment refueling (if it is your responsibility). |
| | 45. Is the site plan refined? Detailed site set down requirements might have to be designed on an hour-by-hour basis for large incoming spares, outgoing spares, and off-line work while fitting everything else in too. |
| | 46. Require MSDS (Material Safety Data Sheets) for all chemical products deemed hazardous - which can be interpreted as almost everything, from all contractors and subcontractors for any products used on site. Safety department to review and reserve the right to bar hazardous materials for which there is a substitute, from the site. |
| | 47. Produce schedule. Distribute at turnaround meeting. |
| | 48. Review all pre-fabs. Look out for excessive numbers of small pre-fabs, extremely large prefabs, prefabs needing weird materials or techniques, prefabs that require certification or special attention due to high pressure, hazardous materials, etc. |
| | 49. Verify that PM work orders are produced and that the PM and PdM inspections are completed. Particular attention should be paid to assets that are planned for service. |
| | 50. Initiate specialized inspections. Include pressure vessels, boilers, piping, tankage, and protection devices. Outside inspectors might have to be called in for these inspections. Work orders may be needed, based on the inspections. Some of the scope for the shutdown will come from the inspection program. |
| | 51. Order NDT (Non-Destructive Testing) on any items on which the scope of work is not yet defined (if it will help). |
| | 52. Order oil samples for all units on which you are considering changing the oil. In some equipment, an oil change is unnecessary. On other units, the analysis will uncover deterioration that should be dealt with. |
| | 53.Gather drawings, wiring diagrams, and shop drawings from sub-contractors, and other specialized information. |

| | |
|---|---|
| | 54. Secure permits or clearances for equipment, including safety, line open, open flame, and confined space entry. |
| | 55. Obtain any needed local building permits. Get state or federal permits for any activities that come under their jurisdiction (if necessary). This is called statutory (under federal, state and local laws) permission. |
| | 56. Review procedure to shutdown the plant from phase 1 (include both operations and maintenance procedures) and see if they still are complete. |
| | 57. Review start up sequence procedure to turn the plant back over to operations (also from Phase 1. Use shutdown protocol as a guide). This protocol should include the sequence to turn on, which processes, and blind list. |
| | 58. Produce a master plan with software. Add revision numbers and MAKE BACK UP COPY |
| | 59. Produce readable copies of all individual job plans. Include work orders in packet for craft supervisors. |
| | 60. Some organizations hang physical tags with work order number on equipment to be worked on to reduce confusion. |
| | 61. Keep adding details to master plan (on software) that were missing from previous reviews. |
| | 62. Draw a resource schedule from above and decide how much in-house resource will be used for the shutdown, how much will be used for other maintenance work and how much will be held in reserve. |
| | 63. Arrange for bulk works (mass rebuilding of valves, steam traps, etc.). Have areas designated for incoming and outgoing items. Inform rebuilder of schedule. |
| | 64. Keep contractor informed about the resource requirements. Rule of thumb for common trades in accessible areas: give the contractor the number of weeks notice equal to the length of the shutdown. In other words give them 4 weeks notice of labor requirements for a 4-week shutdown. If the skills needed are very specialized, or the location is remote, give the contractor (significant) additional time. |
| | 65. Review resumes of contract supervisors and workers. Refuse people without appropriate experience or who have been troublesome in the past. Consider a personal interview and consider taking photographs. |
| | 66. Check your vendors for stock levels of general supplies (everything from welding rods to oil dry and everything in between) and inform them of the shutdown dates. |
| | 67. See that all work orders are activated and that all planning and job packages with estimates, bills of materials, lift paths, and job safety analysis (JSA) are completed. As mentioned this may be the contractor's responsibility. |
| | 68. Be sure plans are in place for noise and dust control especially if the shutdown location is near the outer perimeter of the organization's property line. |
| | 69. Depending on the size, scope and duration of the shutdown, decide on the best method of managing incoming parts. Consider bagging and tagging parts with work order numbers for smaller parts. The logistics team has to decide to use the existing storeroom and existing storeroom personnel or build a new one and staff it. This function can also be outsourced. |
| | 70. The shutdown narrative report is getting pretty fat with all the comments from the planning team. Reports, minutes, displays, contracts are filed in the report. |

By this time you have put into place a physical structure (plans and schedules) and a personnel chain of command to manage the project. Everything is ready to go. You've made your list and checked it twice. Experts in each area, contractors, engineers and supervisors have now reviewed this version of the turnaround schedule.

The large jobs have been broken into smaller projects. Details are being added to fill-in the dark areas on the schedule. These tasks and dependencies should be entered into Project Management software.

**Channels are in place to evaluate added work and emergent work.**

# Shutdown Essentials

# CHAPTER 20

# CONTRACTORS: HOW TO INTEGRATE EXTERNAL ORGANIZATONS

Organizations that have shutdowns have many different relationships with contractors. In some firms the contractor is a small part of the effort, either supplying extra labor or specialized expertise of one type or another. In others, the whole shutdown is designed around a contractor's strong suit. This book has assumed that all preparation and planning is conducted by company personnel but in fact, in many shutdowns, most of the Phase two work is actually done by the contractor.

Some firms bring in a contractor at the initiation phase and the added expertise is invaluable in keeping expectations in line with reality. Sometimes contractors are more effective than company personnel because they are viewed as outsiders with some objectivity. Many firms hire the contractor to prepare the individual job packages with Job Safety Analysis. These contractors then offer a fixed price bid for the package of work they have developed. Beyond your company rules and some legal requirements you can design a contractor client relationship that suits you.

Some shutdown organizations love to hate their contractor. While they're been in a relationship for years, it seems that the contractor gets blamed for everything that goes wrong. Yet the next time the work rolls around, the same contractor is the low bidder and gets the work. This scapegoat role is a convenient one for the contractor to play. It protects the organization from having the unpleasant task of confronting its own shortcomings. While the organization is are blaming the contractor it can avoid looking closely at what practices are in place that don't work or what personnel they have in jobs who are not competent.

Sometimes a long-standing relationship has to be split up. As in any field of human endeavor there are bad contractors as there are bad spouses. In most associations with contractors, as with spouses, there are bad relationships and relationships that went wrong and cannot be fixed. The question is how to avoid the problem in the first place.

One potential proactive solution was offered. Anas Aziz is a manager for Wong Heng Engineering, a major shutdown contractor in Malaysia. He has been involved in shutdowns where the customer sent the shutdown team (both contractors and employees) off site for a team-building workshop. He felt that the time spent was valuable since some level of trust was built up.

It is a challenge to get contractors integrated into the internal workings of the shutdown. Perhaps the team-building workshop that Mr. Aziz described is a good idea. But the two groups seem to have a built in conflict. On the surface it seems that the goals are far apart. The contractor is trying to make money by renting you people and providing you with expertise. You want to minimize costs and the contractor cost is one of the largest.

Standing over this whole battle is the 800-pound guerrilla, the downtime costs. Most of the time the cost of having the plant down dwarfs the separate cost of the shutdown or any savings from the lowest bidder. Of course, if the contractors were comparable then any money saved is real money. The problem is that the best contractor for the job might be the second or third lowest bidder. Thus, taking the lowest bid (which might be mandated by law for some organizations) is often penny wise but dollar foolish.

Then there is the cultural problem of the role of outsiders. In some organizations it runs against the grain to let outsiders to become too close. Generally it is to your advantage to keep the contractor in the communications loop so that they can apply their problem-solving abilities and their experience as members of the team. In fact, experienced shutdown managers such as Larry Jones from Duke Power require their contractors to solve problems (they won't let problems be brought to the shutdown review meetings without the solutions also being produced).

**There are several types of contractors. Each has different strengths:**

**Design and Build Contractors:** These firms are at the top of the construction food chain. They combine the ability of a general contractor with those of an architectural and an engineering house. Specialists (like the ones that build power plants or warehouses) might have significant expertise in that kind of plant or building.

These organizations can be both deep and quite large. They also can generally provide everyone from the shutdown manager on down. The multi-national firms (like Bechtel, Kellogg, Brown and Root, etc) fall into this category. In some situations they will take on responsibility for scoping, planning, validating, and execution on a turnkey basis.

The big international firms will build a plant, put in a town for the workers, and do just about everything in between. Some firms specialize in shutdown work and may move from one shutdown to another. The automotive industry, for example, has firms that specialize in shutdowns for annual model changes.

**General Contractors** usually have some people of their own but might sub-contract, go to the Union hall, or use Agency shops to fill in. They might hire for longer construction projects. All the design and build firms are general contractors (and can fill that role), but not all general contractors have the depth of expertise to take complete charge of a shutdown.

These contractors usually have significant project management expertise and may have expertise in your specific type of shutdown. They also can sometimes supply everyone from the shutdown manager on down. General contractors can span the size range from small local firms to large multinationals.

**Project Managers** function more in the roles of engineers/architects. They only manage the shutdown, making the owner the general contractor. The major firms usually offer this service but any of the types of contractors might take on such a role (particularly if they have more people than work). The compensation might be on a per diem basis or a percentage of the job. A good project manager should have significant expertise in the kind of work and kind of shutdown involved.

**Trade contractors.** These people are mechanical contractors, electrical contractors, etc. that are dedicated to a specific trade. They serve as sub-contractors to the majors mentioned above, usually having a core crew which they supplement with Union Hall and Agency people. Such contractors usually are quite skilled in their trade but it is wise to check that they have experience in your industry and your kind of shutdown. Almost all larger trade contractors will serve as general contractors if the job involves mostly their trade, and other work is incidental. Generally you do not look to them for project management expertise unless the project is overwhelmingly within their narrow band of expertise.

**Specialized Contractors.** These firms are the myriad of contractors for everything from Infrared Scanning or Terrazzo Refinishing. In the oil refinery business there are specialists in catalyst removal, having specially designed vacuum trucks, special techniques, and trained personnel. As such they can be extremely efficient when they are limited to their area of expertise. Usually they are smaller companies with significant expertise in very specific areas. They can be a great help in their specific areas.

**OEMs with installation or repair arms.** These firms are specialized contractors who supply, install, and service their own equipment. They too, have significant expertise in a very specific area. The turbine manufacturers have installed and serviced their equipment in all situations. The personnel travel worldwide, installing, servicing, rebuilding, and solving turbine problems. Backing them up are databases of information about the performance of their equipment. In some high-risk situations or for intractable problems these contractors are sometimes the best choice (they may be expensive in dollars per hour but cheap in terms of downtime and first time repair rate).

## Contracts and legal relationships

You can design the relationship with your contractor to suit your needs for a specific shutdown. For example, where the scope of work is well specified you can ask the contractor to give you a single price for the whole shutdown from planning to execution to closeout. In another shutdown you might just want 150 electricians and 50 fitters with 15 supervisors for 3 weeks. In still another shutdown you might want to rent a shutdown planning team with a manager and subcontract the work to still other firms.

## Types of contracts used for shutdowns

- Lump sum or fixed price - you give a scope of work to the contractor and he/she gives you a fixed bid. Be careful of contractors low-balling a fixed price contract. They may charge you for every nail and screw not specifically called for in the scope of work.

- Unit price - this contract can be expressed in cost per labor hour, cost per square foot of flooring, or an item that can be expressed as a unit of measure. The tunnels in the Hoover Dam were bid by the cubic yard of material removed. The revenue yards were distinguished from non-revenue yards and from contractor mistakes or over excavation.

- Cost reimbursable with a ceiling - you pay all contractor costs including overhead and profit up to a maximum. Usually protected by the phrase "cost not to exceed $X."

- Cost plus fixed fee or cost plus percentage - you pay the contractor's costs plus a percentage or a lump sum to cover the contractor's overhead and profit. In a shutdown and rebuilding contract the contractor added 13% to the cost of any material that the owner requested to cover the overhead and profit.

- Construction management - you pay a percentage of the job and pay the subcontractors yourself. Architects frequently work this way on buildings. Construction manager approves the work and the bills. If you need specific management people you might agree to a per diem charge. Then you pay $1000 a day plus living expenses and travel costs (or whatever amount is agreed) for a manager for the duration needed.

**Some common contract terms**

*Addenda* - Changes made to the contract after bidding but before bids are received. They might be changes to dates, added drawings, or answers to questions that came up in the pre-bid meetings.

*Incentive- Penalties* - Bonuses are paid for meeting milestones and penalties for missing them. As mentioned, be sure you use the right kind of incentives. You might find yourself paying both an incentive for early completion and also paying for rework of their rushed work. Any of the contract types can include these features

*Stipulation* - the contractor agrees to changes to the contract made after an award but before the contractor starts work.

*Change order* - changes to the scope of work made after the contractor starts work. The change order is a written request from the owner that will have a cost associated with it. If the change is massive, then a supplementary agreement is advisable.

*Supplementary agreement* - is a large change to the original contract (maybe a complete rewrite) and must be signed by both parties.

## The story of Contracting

It is best to start the contractor story at the beginning. Getting the contracts signed is the last stage of a potentially lengthy process. Procurement of significant amounts of labor, parts and labor, or design and build services, is preceded by significant amounts of preparation. If money is no object it doesn't have to be this way. In the Manhattan project the major contractors signed on before they even knew what they were bidding on. (Allis-Chalmers signed on to build 5000 compressors before they knew the specifications)!

In the PMBOK Guide (in resources section) procurement management is divided into six areas or phases:

- Procurement planning - what to buy, when to buy it.
- Solicitation management - Preparing the bidder list, getting bids, quotes, proposals, as appropriate
- Sourcing - Choosing who to use
- Contract administration - Managing the contract after it is signed.
- Contract closeout - Completion of the work and settlement of the contract including payments and resolution of all outstanding items

## Steps of Contracting

In one instance, the cost to prepare the documents and follow through on the process exceeded $250,000 (in addition to the cost of the shutdown!). The organization had to hire experts to listen to the different people in the company and define and formulate the problem scope, determine the depth of the problems, write the specifications, and review the responses. The team also thought through many different scenarios and tried to word the contract to avoid any problems they could think of.

Team members also had to attend all the pre-bid meetings. The process took about a year, with people from different disciplines working on it (including a whole team from the owner, but not all full time, of course).

Not all contractor procurements are so complicated. Some have only a few steps. If the contractor is also responsible for engineering there might be a few more steps. Some of the steps (after the organization has agreed to the scope of work) might be:

- Preparing the specifications and drawings for the bid, including design if needed (if design is included in the bid then the specifications will show what is to be designed in detail). Define work to be contracted, based on the work orders and job packages. The better the definition at this stage the better the job will go. This definition is another word for scope of work. Make sure the scope also includes special conditions (such as working 200 feet off the ground; special requirements (code welding); built-in non-productive time (for instance, security access into an airport). The contractor should be aware of safety training requirements, PPE, and site rules, as they impact the contract. The contractor then bids on the scope of work usually as expressed by drawings, specifications, or both.

The second technique commonly used on outages is to hire a given number of qualified trades-people from a contractor on a time and materials basis. They can be hired with and without supervision. This approach is common where the outage consists of many smaller jobs and the contractors have worked in the facility many times before.

**On many outages both methods are used at the same time.**

- Call a pre-bid meeting where you communicate your ideas to the potential contractors (do your best to make sure they understand). Discuss the quality of materials needed (if the contractor is to be the supplier). Discuss all the ancillary issues such as PPE, safety training, site rules, security of site, storage, bathrooms, etc. Ask attendees at this meeting to submit questions.
- Search for potential contractors. Where possible, use known, high quality, contractors. Advertising is appropriate for large jobs. The job engineer or architect will also have lists of contractors they have used in the past. If you have periodic outages you will have a list of contractors that served you well from past jobs. It is always a good idea to invite smaller contractors to take on appropriate tasks, to see how they perform and to groom them for bigger projects in the future.

- Request information about the capabilities (engineering, management, financial, reputation, etc.) of the bidders for pre-qualification. On larger jobs, check out the contractor's finances, credit, insurance, and staff. Visit other jobs to see their quality, and call references.
- Hold pre-qualification meetings
- Make pre-qualification lists
- Get answers to pre-bid meeting questions sent with Request for bid or request for proposal.
- Send out the RFB (Request for Bid), RFQ (Request for quotation) or RFP (request for Proposal) to those bidders who made the qualification cut.
- Hold an official opening of the bids
- Analyze the bids
- Award the bid and negotiate the contract. Negotiate to improve areas where the contractor's bid is inadequate. Be as complete as possible about responsibilities, who supplies what, who supplies housing and transportation, where to unload, rules, your contact, who they interface with, clean-up, security, keys, when (and amounts of) payments will be made, etc.
- Mobilization of resources involves setting up of schedules for mobilization, commencement, and completion. Agree on a reasonable schedule and measurable milestones, also on safety briefing, standards, and PPE. The agreement should include who is responsible for government permits and shop drawings, or detailed plans.
- Investigate computer interfaces. The best situation is when the contractor and the owner use the same database and the same package (and same revisions) for work management and project management.
- Provide lead time to build, pack and deliver (if some part of the job is to be built off-site)
- Prepare the area to be worked on. If possible, empty the area to avoid breakage.
- Manage the contractor. Keep complete records and provide feedback. Conduct frequent inspections. Problems should be identified as early as possible. Require paid receipts to prove subcontractors and material vendors have been paid.
- Insure you do nothing to interfere with the contractor's efficient work, and issue permits efficiently and quickly.

- Specify that all change orders are to be approved in writing. Sign off on completed work as soon as possible.
- Administer the contract. Resolve conflicts and problems quickly. Keep very careful records of the whole process. Monitor the contractor to be sure that contract requirements are fulfilled.
- Evaluate each contractor on a regular basis for quality, service, cost, and fulfillment of contract terms. Afterward, write up a short narrative to put into the file describing how the job went, and file a copy in the project narritive.

The team wants the best contractor the company can afford. In the maintenance world the cheapest is often not the best, but many organizations are obligated to award to the lowest bidder so the qualification process is necessarily strict. Saving a few thousand dollars on a shutdown might cost you hundreds of thousands in extra operational costs, not to mention change orders and extra work for the owner.

Sometimes it is an easy choice since the organization uses the same two or three contractors over and over. When you are evaluating contractors for the first time, the challenge is choosing which of the responses would provide the highest probability of success for your kind of project.

### Get them on the inside

Contractors need to be a true part of the team. As soon as they are on board they should be included in all relevant meetings, E-mails, and document distribution lists. The more they feel inside, the more likely they will act as a member of the team.

In specialized industries such as pharmaceutical manufacturing, the contractor has to know a great deal beyond the trade to be effective. To bring the contractor inside the shutdown process requires them to have a significant amount of knowledge. For example, the procedures for closing out jobs and documentation in some circumstances, take longer then the job itself. When this happens, you will have to cultivate contractors constantly.

Mr. Lim Hong Khoon of Schering-Plough in Singapore said they have a few contractors who are familiar with the rules and get most of the work. He is always looking for more. He said, when the island's big contractors get low on work they might bid one of his jobs. Usually they bid one, but after losing money because of the special requirements they never bid again. His contractors have to submit detailed job plans, job safety analyses and lift plans (for any substantive lifts). Practice dry runs might also be required. He said they want the execution to be perfect with no surprises.

## Be a smart customer

As you choose your contractors, make sure they can meet their commitments. Look at the depth of their organization, the time of year (is it a busy time even without your job?). All it takes is for the vendor to have a string of proposal wins and your shutdown can be caught short.

**Idea for action:** Request resumes for at least the key players in the shutdown if not for everyone that will be on the site. If resumes are not forthcoming, then simple summaries of experience and qualifications may suffice. Robert Schultz, Head of Shutdowns for RAS Laffan LNG in Qatar, personally interviews the key contractor supervisors before he is satisfied.

Mr. Schultz has another trick to make his job easier. He requires that the contractor add into their bid an area supervisor (instead of, or if they want to, in addition to the more common trade supervisors). The aim is to insure there is a responsible party in each area where the shutdown is working. He said that in the past, the craft supervisors would point fingers at each other when there were problems. With this arrangement he has a contractor area manager to hold accountable.

When you go to this much trouble there must be sanctions for late substitutions. Some contracts call for deductions when there are substitutions for approved people. To forestall substitution games, one company takes pictures to be able to identify the contractor people they interview and approve.

You have a responsibility in this matter. Be a good customer. Be prepared for the contractor to come on-site. Don't negotiate deadlines, milestones, and budgets that you suspect cannot be met. If you are an important client, many contractors will typically go to great lengths to make a sale, and then fall on their face.

**Supervision:** Specify the number and coverage of the contractor's supervisors in your bid. Some contracts even have deduction clauses for inadequacy of supervisor coverage.

**Mobilization and demobilization:** have the contractor draw up a mobilization plan. Give the contractor enough time to mobilize adequately (or you may be stuck with whomever they can find on short notice). See that the plan allows you to vet the key personnel and has adequate time built in for site training requirements. One of the most difficult issues is mass safety training for contractor personnel. Robert Schultz, mentioned above, has a network of authorized training vendors that he uses to provide the mass training. In the past, his training department quickly became overwhelmed with the job of training 500 contractor employees within the short mobilization period.

At the end of the project there should be an agreement about how demobilization will take place. You might want to debrief certain members of the contractor team before they disappear to the next job.

**Subcontractors:** be aware of all the subcontractors (subs) who will be on the job. There should be a record of what subs will be bidding on what parts of the whole package and after the award you'll know which sub will be doing what work, where, and with whom. You should have the right to bar a sub from the site if they have caused problems in the past. Of course, have a mechanism in place to verify that subs are being paid (this is a standard operating procedure for construction jobs).

**Incentives:** if it is in your best interests, offer incentives for on-time completion, safety, and achieving quality standards. Be careful to incentivize the right things or you'll find work was finished early, but with accidents and start up problems.

**Contractor Communications:** with whom will you communicate within the contractor's organization? Most contractors want a single point of contact. The shortcoming is that the point of contact may become a bottleneck for information and have trouble being up to date on everything going on. Work this out with the contractor ahead of time. You appreciate the single point of contact; they should appreciate your need for timely information.

## Tips to Avoid Claims - Want to avoid problems with your contractors?

### Follow this advice...

1. Avoid loose specifications (on both materials and work to be done) on bid contracts. Too loose a work scope can be a symptom of inadequate time or bad planning on T&M jobs.
2. Don't always take the lowest bid! If you are required to take

low bids, have a complete pre-qualification process.

3. Keep a fair and complete set of contract documents and review them periodically. Also, keep good records and avoid sloppy record keeping.
4. The whole point of this course is to avoid haphazard planning (have a functional planned schedule and manage from it). Agree on the schedule with the contractor.
5. Don't accept the contractor's standard contract form unless your lawyer says it's okay. The contractor would want to use a form that would favor them in disputes.
6. Agree on a definition of performance, so that everyone knows and can define what a good job would look like. Include clauses like "all work is to be done in a professional and workmanship manner." "All work will be in compliance with applicable electrical codes." Also make reference to well-recognized quality standards from outside organizations such as ANSI and ASTM.
7. Be clear about how the site is to be left at the end of each workday (if it is not a 24/7 shutdown). Who will be responsible for locking up, cleaning, and debris removal? Be particularly careful of areas where other employees must pass on the way to their workplace, or areas where the public comes near the site.
8. Depending on the agreement, have deduction clauses and spell out what you will charge back and when you will charge it. Examples would be debris removal, clean up, incomplete work, inadequate work, missing firm completion dates.
9. On T & M jobs clock the contractor's workers in and out of the site to verify hourly charges. In any shutdown, the contractor's employees should record time on the work orders for your information and records.
10. On construction jobs with a general contractor verify that material vendors and subcontractors have been paid. You could have paid off the general contractor and still be hit with liens.
11. Verify insurance, and have an agreement about what happens when the contractor damages your property (or worse, damages a neighbor's property). Include pavement damage, fence damage, and damage (either accidental or purposeful),

from contractor employees. Require a certificate of insurance covering general liability, property damage, workmen's compensation, and auto liability.

12. Aim for early resolution of disputes. Require immediate resolution of disputes, both on your side and theirs.
13. You should spell out a way to fire any contractor. You need to define how and why you can cancel the contract. Otherwise you might find yourself with a mechanic's lien over an inadequate job, or having to pay for the entire contract but only getting half the work.
14. If it is a fixed price contract, you should have a schedule of extras. A common ploy is to low-ball the bid to get the job, and then flood the company with small extras. Look for clauses like "all extras not included in the original price must be agreed in writing, prior to the commencement of the work."

## Disadvantages of contracting

Of course there are downsides to using contractors. For example, the downsides might include losing control of the shutdown because the contractor is responding but not communicating.

There are some serous pitfalls when the work scope is not complete or not well defined. You could open yourself up to excessive charges and have no recourse. Your scope is your protection against overcharges. This potential for loss is a problem for both in-house crews and contractor crews, but it is more expensive with contractors.

If the contractor has had a lean time they might take jobs outside their core expertise just to keep people busy and pay their bills. They may be stretching their expertise just to get the work. The result for you is poor quality work and possibly excessive labor charges. Fortunately most contractors want to get additional jobs and look good, so they limit themselves to work in which they have experience.

It's also sometimes hard to get the last part of the job completed. The bulk of the contractor's people are gone and the ones left are preoccupied about getting on with their next job. We've also seen contractors who were satisfactory at their trades but seemed apathetic about compliance with safety rules and protecting their employees. You also could get off on the wrong foot and have a contractor-owner shouting contest. A contractor can miss deadlines, disrupting the schedules of other contractors, which cascades through the schedule.

**Reasons to use a contractor during an outage:**

If contracting is so full of pitfalls, why do we even bother hiring contractors for shutdowns? The simple answer is, if organizations could do shutdowns without contractors they probably would. There are specific reasons that usually can be traced back to the fact that proper use of contractors saves money.

Some recommendations:

1. Reduce the elapsed time of outage
2. You don't have enough people to do it yourself (even if you wanted to).
3. Improve quality by hiring specialists who have long experience in the area.
4. You lack in-house skills (heavy rigging, tank construction, etc.).
5. You don't want to manage the job (you are hiring the contractor to do that).
6. You don't want to lose control of existing work to make room for the shutdown.
7. You lack the appropriate license (even if you have the skills).
8. You lack the specialized equipment, and job specific chemicals, or materials.
9. Work that can be moved off-site should be because it will not strain your resources or infrastructure.
10. Liability can be reduced by having another insurance company (the contractor's), hold you harmless from damages.
11. You can reduce hazards to your own employees (in tank cleaning, asbestos removal).

In one interesting case the contractor's flexibility saved the day. It seems that headquarters planned a critical shutdown during an important religious festival. That religious group happened to dominate the contractor's usual crew, and it looked as if the shutdown was going to be in trouble. The contractor pulled the rabbit out of the hat when he realized he had another crew being mobilized for another job that was dominated by a different religious group. The mobilization allowed some people to be slipped out for a limited period. The second crew was used to keep the work moving for the week of the festival.

## Changes will occur

Changes to the work scope (change orders) are a popular place for contractors to make a profit. When a change order is issued, it isn't usual to bid it out and the contractor's rates are accepted. Hopefully your contractor has specific language about charge rates. Whatever you do, change orders always seem to be expensive.

In most shutdowns, scope creep is inevitable because unknowns existed when the shutdown started. Vigilance must be maintained. Some emergent work can be accommodated by your own employees and not impact the contractor. Other work can be covered with contingency funds built into the budget for unexpected surprises.

The key is to protect the overall project, protect your boss from looking bad, and protect your team from disruptions to the priorities that changes create. After everything is said and done, manage the changes. If possible get an estimate from the contractor.

Business priorities change and some requests for changes won't go away. Formalize the process of accepting changes. Create a change order process at the very beginning of the project. As mentioned it should be part of the initial contract. How many times have you heard, or worse yet, thought, "While we are doing this thing A, it would be silly not to do this other thing B?"

Within reason, fight the urge. If you do take on the extra work, make everyone up the ladder sign off on the change order, and make sure they are informed in explicit terms on the consequences of the change. Adjust the remaining duration and cost estimates when scope changes are approved.

CHAPTER 21

# ACCOUNTING, COSTS, AND BUDGETS

When all the planning is completed and the work list finalized, the estimates for labor, materials, and equipment is converted to a Shutdown budget. This budget becomes the cost baseline for the shutdown.

The baseline becomes part of the tool to help control the shutdown event and insure maximal utilization of the organization's funds. The baseline also helps by providing navigational aids to the shutdown management, showing when the project in trouble (for instance, by spending money too quickly or too slowly).

One major player in the budget for a shutdown is the location and conditions of the workplace. For example, oil wells are drilled under a variety of conditions. There are oil wells even in populated areas (there used to be many wells in Los Angeles County), in rural areas remote from supply lines (such as Prudhoe Bay, Alaska) and off shore. Each area has a factor that is associated with the cost of shutdowns. For example the same shutdown off shore might have a budget 10 times higher than the same work on shore.

Most organizations will assign a master code or account number to each particular shutdown. All work, materials, equipment rental, and any other purchased service or item, would be charged to this code or number.

To get more specific costing information it is necessary to do more work on the front end and during the shutdown. All jobs and all their charges should be recorded on standard company work orders. The estimates must be complete and accurate for the budget to be meaningful. Most organizations that use CMMS already have links built from the CMMS to other systems. The work breakdown structure is a handy way to total costs by section of the shutdown.

**Inputs to accounting (Budget):** Work lists with estimates, resource lists, contracts, purchase orders, schedules of when events will happen, change orders, extra work, and emergent work.

**(Actual):** Invoices (matched to packing slips or time cards), completed (and filled out) work orders.

**Outputs from accounting:** Revised cost estimates, updates to budget, estimate at completion (one of several kinds of projections of costs based on sunk costs and work left to complete).

**Accurate accounting**: Accurate accounting is essential because so much of the maintenance budget (30-70%) is consumed by shutdowns.

**Accurate shutdown accounting is made more difficult because:**

1. The shutdown event is compressed in time and intensity so that information about proper charging floods in, and if it is not captured correctly, might be forgotten or lost.
2. Under the best circumstances work orders are difficult to read and interpret. Under shutdown circumstances they might be impossible to understand (without a powerful system that forces the mechanic and supervisor to completely and clearly fill out every field).
3. Contractors are used and they don't always submit bills in a timely way or in a format that is easy to relate to your accounting categories.
4. Bills for rental equipment might not indicate that they were used for a shutdown.
5. Some maintenance work uses exactly the same resources as shutdown work, making it difficult to separate normal maintenance costs from shutdown costs (particularly if it is a partial shutdown and other maintenance work is still going on).
6. There is sometimes an incentive or a temptation to transfer costs out of the shutdown to routine maintenance (to make a budget number and look good), or out of routine maintenance into the shutdown (to make a different budget number and different people look good).
7. The cost to accumulate accurate costs (by work order for example) might be prohibitive.

8. Shutdowns are on the cutting edge of issues of capitalization and expensing for accounting and tax purposes, which might make some artificial distinctions necessary and complicated to manage. The decisions on these issues are unrelated to the maintenance and engineering of the shutdown and are entirely driven by tax law and accounting goals.

## The Budget spreadsheet

One tactic is to prepare a spreadsheet with all the major projects and work orders down the rows and the expense categories across the columns. The rows can be keyed off work order numbers or WBS (Work Breakdown System) codes, or both.

There is some advantage in including the WBS codes (if you are already using them for project management). WBS codes will allow costs to be tracked against individual work orders or activities where that is practical (at a terminal activity level). Where that is not practical, the costs can be charged to higher levels of the WBS. Also, the WBS facilitates totaling of costs to higher levels.

For example, you might rent a crane for a whole project such as 'Install new stack' that might have 15 work orders. You could charge the crane against the whole project (very simple), rather than trying to allocate it by hours or another method (very complicated), to the individual jobs making up the project. Charging the crane to the whole shutdown is another possibility, but does not capture the costs for installing the stack (if that was the only work for which the crane was used.)

Constants will be charge rates (the hourly salary + benefits + overheads) for each craft (many places might have only two or three charge rates) such as C1 and C2 below.

| C1 = $50 / hour. C2 = $40 / hour | | | | | | | | | |
|---|---|---|---|---|---|---|---|---|---|
| **Work Order Number** | **WBS** | **Short description** | **Hours C1** | **Charges for C1** | **Hours C2** | **Charges for C2** | **$ For parts, mat'l** | **Equip-rent** | **Etc.** |
| **65291** | 3.1 | Replace stack | 200 | $10,000 | 50 | 2000 | $4500 | $500 | |
| **72359** | 4.1 | R & R support Brackets X4 | 1000 | $50,000 | | | $25,000 | | |
| **72360** | 4.2 | R & R Support brackets X5 | 800 | $40,000 | | | $18,000 | | |
| **P218** | 5.0 | Install new precip. | 75 | $3750 | 150 | $6000 | $125,000 | $9500 | |

The Etc. column represents as many columns as are necessary for all other resources for each job. There might be several versions of this spreadsheet over time with jobs added and others deleted.

After the work orders are listed the spreadsheet would add all the columns and have lines for extra and emergent work. The Extra column might add 5% of the sum of all the work orders. The Emergent column might add 15% of the sum of the work orders. Your historical percentages would be useful (assuming you have tracked emergent and added work from shutdowns of this type in the past).

After Extra and Emergent would be a series of charges for site-wide requirements not charged to individual jobs. These jobs would include everything from spare cranes to eyewash stations, from welding tanks to extra site security, from free access items to extra PPE. This spreadsheet becomes a control document once the shutdown begins. Every shutdown bill would be coded to this spreadsheet.

| Work Order Number | WBS | Short description | Hours C1 | Charges for C1 | Hours C2 | Charges for C2 | $ For parts, mat'l | Equip-rent | Etc. |
|---|---|---|---|---|---|---|---|---|---|
| Extra | .2 | Extra work orders (authorized) | 2000 | $100,000 | | | | | |
| Emerg | .3 | Work emerged during shutdown | 6000 | $300,000 | | | | | |
| Site 1 | .4 | Site wide service #1 | | | 2000 | $80,000 | | | |
| Site 2 | .5 | Site wide service #2 | | | | | | $1000 | |

This spreadsheet accumulates the costs for the shutdown. Intermediate versions should be saved in the project archives.

### How to use the budget spreadsheet

It is very useful to take all the estimates from the work list and expand the costs into the budget spreadsheet. Estimate all the site way costs, emergent work, and other unknowns. This first version of the budget spreadsheet is the starting budget. Save this record, and don't change it once it is approved. Graphs are generated showing the amount of hours and dollars of various categories.

Clear all the numbers from the spreadsheet to make a blank sheet with titles and formulas only. As soon as the shutdown begins, route all information through the person responsible for this spreadsheet. Add in data about actual costs as it becomes available. Estimates, and rounded numbers are okay. Every week at the shutdown update meeting (or at milestones), rerun the spreadsheet graphs for comparison with the budget (called the baseline).

Along with an estimate of work to be completed, this spreadsheet will give you a good leading indicator of budget performance at the end of the shutdown.

**Keeping the books: Chart of account approach to shutdown bookkeeping**

Create a chart of accounts for each shutdown. Here are some of the possible accounts.

**Materials:**
Maintenance materials
Large spares
Maintenance parts
Maintenance supplies
PPE (disposable)
Inside prefabrication
Outside prefabrication
Rebuilding services

**Rentals:**
Equipment rental
Tool rental
Vehicle rental
Demurrage
(welding tanks, rail cars, etc)
Office trailers, toilets
Mobile accommodations
Office equipment rental

**Contracts:**
Contracts (fixed price)
Contracts (cost plus)
Contracts (unit priced)
Project Management services
Service contracts
Contract labor
Contractor travel,
accommodations, per diem
Engineering contracts
Inspection services
Other Consultants
Training contracts for inside
and outside labor

**Shutdown overhead and other costs**
Security
Housekeeping services
related to Shutdown
Waste removal (both
hazardous and
Non-hazardous)
Project Management support
Shutdown management team
Damage Repairs
Special Insurance
Legal costs
Accounting costs and G & A

**Labor (inside):**

In-side Maintenance labor straight time (with fringes)
Overtime
Purchasing and accounting support
Inside Maintenance labor - overtime
Borrowed labor from operations
Borrowed labor from other plants

CPM/PERT is also a primary budgetary development tool because budgets are associated with each activity and event of the network. Furthermore, CPM is a budgetary control and project management tool throughout project execution because it provides alarms when:

- Interim targets are missed (earliest event time on the critical path)
- Float is reduced (latest event time off the critical path)
- The rate of financial commitment at any position in the network exceeds the budget

## Spending

This accompanying chart is the spending pattern for a small shutdown. The costs start slowly and increase at an increasing rate and then decrease toward the end of the shutdown. If this pattern is charted the result is the curve Cumulative Cost Line Graph. By developing a graph like this for a proposed shutdown you'll be able to see and predict what the costs should be against what they are for each reporting milestone.

| Period | Amount Spent in Period | Cumulative amount spent |
|---|---|---|
| 1 | $7000 | $7000 |
| 2 | $16,000 | $23,000 |
| 3 | $16,000 | $39,000 |
| 4 | $35,000 | $74,000 |
| 5 | $54,000 | $128,000 |
| 6 | $51,000 | $179,000 |
| 7 | $16,000 | $195,000 |
| 8 | $10,000 | $205,000 |
| 9 | $13,000 | $218,000 |
| 10 | $12,000 | $230,000 |

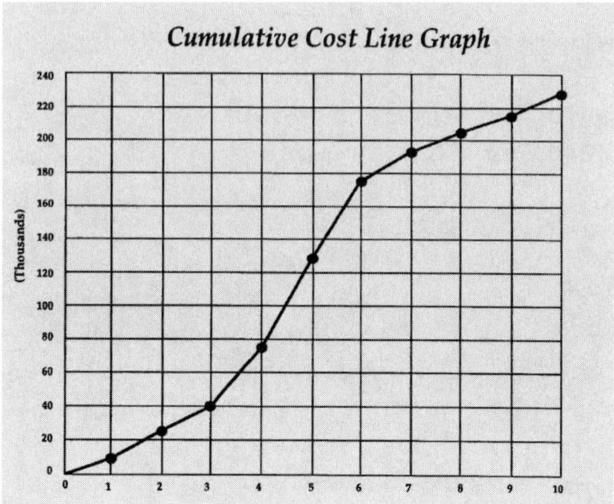


Graph of cost date from table on previous page.

# CHAPTER 22

# LOGISTICS: ORGANIZATION FOR THE PARTS, MATERIALS AND SUPPLIES

## Materials Management

A typical shutdown might have 1000 or more parts and materials from a hundred vendors to manage. Initially, the purchasing team leader, and eventually the logistics leader, are key players. They are responsible for the process from ordering, shipping, expediting, delivery, storage, on-site management and tracking, issuing and/or delivery to the work site. The whole chain starts with the planner, who calls out the part needed on the individual job plan Bill of Material (BOM).

Managing delivery of all the parts is an essential and frequently a massive job. On a small shutdown you even use a PIM (Personal Information manager - such as the schedule in Outlook), to remind you when parts are due. On a typical shutdown the major parts (at least) should be set up in the schedule as activities so that they are managed along with everything else. The duration of the activity would equal the lead-time of the part.

### Far, far away

In a major city, standard bearings, seals, motors, drill bits and a wide range of shutdown parts, consumables, supplies and small tools can be obtained in a few hours. North of the Arctic Circle, in the middle of the ocean, or on a small island without a large industrial base, the supply situation is quite different. Without industrial distributors, courier services, frequent airfreight service, and other supply lines, each purchase can be a major undertaking.

The material management function takes on a much greater significance when plants are located far from these channels of supply. The material management plan has to be increasingly bulletproof, accurate and reliable.

International sourcing is another tough area for all shutdowns but particularly for remote sites. Differences in language, time zone, and attitude, make business more challenging. Fortunately the Internet is starting to simplify international sourcing, but the part or supply item still has to get to your location. Many smaller vendors are still reluctant to do business overseas, and maintain a parochial point of view.

### Master Material List

Consider a master material, spares, and supply list (you can even add tools and equipment or keep a separate list) of the type:

| Item | Work order number | Date due | Date ordered | PO # | Responsible party | Status |
|---|---|---|---|---|---|---|
| W1281 Clamp | 28342, 27012, 27000 | xx/xx/xx | xx/xx/xx | SD1726 | Heffner | On order |
| T66 Starter | 31789 | xx/xx/xx | xx/xx/xx | SD934 | | |
| Omega 4789 | 28340, 28341 | xx/xx/xx | xx/xx/xx | STOCK | Mathews | Stockroom |

If the list was kept in a spreadsheet it could be sorted by date due for follow-up. To follow-up, first check with receiving and be sure the item has not come in recently. Call the supplier to find out the status. If the item has been shipped, get the tracking numbers and trace the shipment. As a last resort, cancel the order and find a new vendor.

## Function of the Logistics Team

The logistics team is the low visibility group (until we can't find something), on which we depend for the smooth running of the shutdown. The mission of material management is to be sure that the right materials and other resources arrive at the individual job sites at the right time and are delivered into the hands of the right shutdown crew (in good condition!).

To put material management in perspective think of a military campaign. All the ordnance, spare parts, food, water, and consumables, must converge on the right place at the right time. If they don't, the kingdom could be lost for want of a nail!

Logistics is responsible for receiving, storage (in a way that will not degrade the material), delivery, and disposal of all materials and tools. In addition to managing the "thing itself" logistics must have an information system to know what the status is right now and at some target time in the future. They have to be able to answer a variety of questions (usually on the spot or in a few minutes)

**Receiving and managing all objects and substances come under logistics**

Accommodations
Bulk materials
Cleaning gear
Delivery
Equipment
Facilities (toilets)
Hazardous materials
Land (plot plan)
Lifting gear
Machines
Parts
PPE
Rail cars
Safety facilities
(eye wash,
decontamination)
Scaffolding
Services
(electric, air, water)
Supplies
Tools
Trailers (storage)
Transportation
Wastes
Welding tanks

**Accountability includes having instant answers to the following questions that come up during the shutdown:**

Where is it?
Has it arrived?
Did it pass inspection?
When will it arrive?
Can you pick it up?
Can you deliver it?
Can you dispose of this?
Did we get rid of it?
Did we return it?

One of the logistic functions is the unloading and movement of large spares. In many operations, the large spare is unloaded as close as practical to where it is being used or moved there immediately. Protection from bumping and damage is of paramount importance. The site plan should reflect a set down area right near where large parts are to be used.

Logistics is a scheduling function that insures critical resources are on site at the right time. Imagine the logistics problem faced by off shore oil drilling rigs. There is no place to put anything so you have to organize parts barges. The people have no place to stay so you have to organize accommodation barges. If you have to lift anything beyond the capacity of the crane you have to float in a bigger crane. You also have to deal with critical parts and spares accidentally falling overboard. All in all, off shore rigs present a pretty complicated logistics problem.

## Managing the spares on the site

Use the existing storeroom: Where there is enough extra space under the control of the storeroom, the storeroom is located near to the shutdown, and you have (or can get) adequate shifts of coverage, using existing supply structures makes sense. Of course large spares are stored where called for on the site set-down plan, developed by the shutdown planners, working with logistics.

If there is not enough room, or you don't want to disrupt the existing operation, a temporary storeroom will have to be designed, built, staffed, and managed by logistics. Elements of logistics can be contracted out (as well as the whole function).

Idea for action: On some jobs the storeroom issue can be simplified by locating storage trailers near key units to be repaired. These trailers can be used to store spares and tools unique to the work order or project. Use of such trailers can sometimes take enough pressure off the storeroom to make its use feasible.

## Be careful of what you assume

A critical function of logistics is to verify that the storeroom has an adequate supply of common items. A quick review of the purchase orders issued just before and during the last few shutdowns will shed light on the kinds of common items used. These items can include rags, gloves, penetrating oil, free access items, wide ranges of nuts and bolts, oil dry, and a host of other items. These items are generally not specifically called out on the Bill of material.

The shutdown rolled around and the manager thought he had every thing well in hand. This chemical plant had periodic outages and availability of supplies had never been a problem. The material planner spot checked the warehouse and reported that they had a couple of 'cases' of welding rods of the type necessary for the material and usage.

One day into the shutdown they ran out of welding rod. The distributor was also out because of a sudden demand from another nearby chemical plant. Welding rods had to be found and flown in. Two days were lost looking for rods and bringing them into the plant.

It turned out the warehouse didn't have a few cases. Instead it had only a few boxes. The cases were almost empty – one box in each. Of course, now the assumptions are more carefully checked. Vendors also are checked for stock levels of general supplies and informed of the shutdown dates.

### Management of conflict

One of the functions of logistics is to verify that the parts called out on the bills of materials for all jobs are reserved (tagged and bagged), in the warehouse or other location. The conflict that arises is that if the plant is not entirely shut down, maintenance might need the part for an emergency repair. If the shutdown job is not for a few weeks, can the part be released? One utility has a rule that the machine that is down gets priority, but this can be a deep question needing to be discussed and agreed to. Perhaps the policy team should have its say. It boils down to the consequences of the breakdown versus the consequences of the work order not being done on the shutdown (if the part cannot be replaced in time).

### Large items

The logistics of large items have to be scheduled and optimized. For example, if a large item comes in, the crane might unload it immediately and deliver it to where it will be needed to avoid a second move.

### Developing and managing the Site plan

The plot plan shows the location of every item important to the success of the shutdown. A good site and logistics plan is as important as a good schedule. The plan lays out where big items will be stored, worked on, and staged for shipment. It shows where the scaffolding should be positioned when the tank is lifted into place and a myriad of other details.

The site plan should start from a large plan drawing of the site, which can be obtained from the Engineering office (hopefully) or from the last shutdown files. The logistics team should field check any drawing and is responsible for making sure the plan is accurate and up to date as to new construction.

The plan should show all buildings, tankage, outdoor process equipment, fences, utilities, and natural features. All roads through the site should be shown, especially public access roads. Some plants have overlays for utilities, piping, aerial wires, and process flows, which would be useful. Areas around the use of hazardous or flammable materials should be highlighted. There will be a lot of work going, so it will be useful for the drawing to show any land outside the fence that can be used for staging or rebuilding. Special issues, such as the height of pipe overpasses and non-weight bearing ground should be noted.

Depending on the level of the hazard, all fire assembly areas, eye wash stations, decontamination units, nurse's stations, fire fighting equipment, utility shutoffs, and any major hazard area, protective area or device should also appear. This overlay would be part of the training for all personnel who come on the site (for any reason).

These existing drawings show the site as it is (without a shutdown). The logistics team then transfers the information about the shutdown to a series of copies of the plot plan. The big areas are done first with temporary storeroom, major equipment, lay down areas, accommodations, and any large storage. The next step is to break down the shutdown's activities on a daily or shift basis to see what major logistical moves are required. This breakdown process can be done on a day-by-day or even hour-by-hour basis (if needed). Crane lift paths, and major spare movements would be considered and planned.

When the site plan is completed it is submitted to the shutdown manager at the appropriate meeting. Other responsible parties then try to think of anything the logistics team missed. When the plan is approved it becomes the official roadmap to logistics management.

One new technology is the ability to post the current plan to the shutdown intranet portal. Teams from around the site can log on and find where necessities are stored. The logistics team can update the plan as needed.

Offshore oilrigs present special problems for logistics. In these special circumstances the site set down plan or the site map is particularly detailed and necessarily complex. There are few places to store parts and little extra space. To keep the parts supplies the rig rents barges for parts storage and staging. The limited accommodations also make it necessary to rent barges with rooms, cafeterias, and social rooms

### Incoming inspection

A shutdown can get stopped if a replacement rotor is damaged or turns out to be the wrong one. It is far better to stop the shutdown while the plant is still running than to take it down and find you can't complete the work.

Incoming inspections should be quick, thorough, and ruthless. Speed is essential so that the team will know quickly that there is a problem with incoming materials. Thorough speaks for itself. The inspectors need to know what they are looking at and what is critical. Ruthless means that if there is a problem it is dealt with efficiently. If the part has to be returned, so be it.

### Incoming spares

All parts are to be received, verified (as matching what was ordered) counted and compared with Purchase Orders. Packing slips should be matched against Purchase Orders. All parts should be inspected, tested (if indicated), shelved, slotted, or stored by work order number. A good idea is to bag and tag by work order, or have plastic totes for each job. One plant bought used bus station luggage lockers. Spares were stored by work order in large plastic tubs and the locker key was dropped into the plastic work jacket for that work order. Disposable bits, moderate sized tools, and consumables can be added to the totes if they are called out for the job

### Before Logistics there is Purchasing

Before we can have a logistics problem we have procurement.

Starting at the beginning of Phase two, procurement is at the forefront of the process. One of the deliverables in Phase 2 is a procurement schedule. A large shutdown will have a complicated procurement schedule that will rival the complexity of the overall project schedule.

According to Mark Goldstein, who has extensively studied this area, there are three major levels to the procurement of parts, materials, supplies and other items that are based on the amount of information that is unknown. Each of these levels has a different cost of acquisition and internal lead-time.

| Type of Procurement | Internal Lead Time | Cost of acquisition | Unknowns |
|---|---|---|---|
| Casual (go to web site, pick some things out and use a purchase card or existing account to pay). | 1-2 days | $10-$25 | What we need-Known<br>Price- Known<br>Vendor- Known<br>Delivery- Unknown |
| 3 quotes- The purchasing agent calls around (or Faxes or E-mails) for prices to known vendors. After a day to a week the PA reviews the quotes and chooses a vendor | 1-2 weeks | $100-$500 | What we need-Known<br>Price- Known<br>Vendor- Known<br>Delivery- Unknown |
| RFP, Bid and Proposal - Engineering prepares drawings and writes specifications, several steps to pre-qualify, answer questions, RFQ or RFP sent, bids received and analyzed, award made | 1 month – 2 Years | $5000-$750,000 | What we need-Known<br>Price- Known<br>Vendor- Known<br>Delivery- Unknown |

Given the lead times for the different types of procurement when should we place orders for all the materials, rental equipment, and contract labor? Contract labor will be discussed in another section but the steps for procurement are the same.

### Internal Lead-time

For major items (such as the fabrication of a special reactor vessel, or supplying a packaged boiler) there are several lead times to consider (some may not apply in all situations).

- Designing the special reactor
- Preparing the bidding specification, including design if needed
- Pre-bid meeting with questions submitted
- Answers to pre-bid meeting questions sent with Request for bid
- Sending out the Request for bids
- Waiting for the bids to come back
- Analyzing the bids
- Awarding the contract
- Lead time to build, pack and deliver

As long as the groundwork is done, small items can be purchased quickly.

## Parts vendor Selection Basics

The issue of Parts Vendors for Shutdown situations is, can you count on what they promise you? A few dollars saved is nothing compared with a missed delivery window. The same vendors should be used for routine purchases as for shutdown purchases. They should be brought into the process and should consider having representatives sitting in on some of the shutdown meetings. If they are in the loop they can intelligently stock up to support the shutdown needs.

If the contract for spares or materials exceeds $50,000 for a single shutdown, and the potential firm is small, the financial status of the organization should be reviewed. This would include a Dun and Bradstreet (credit) report, trade references, bank references, financial situation of owner (or annual report if public). In short, does the firm have the financial clout to support you and complete the shutdown work?

Review the proposed terms of sale in detail. What is the published return policy, stock balancing policy, warranty policy, and shipping/delivery/pick-up policy? Are there any incentive plans (plus any informal deals they have offered your firm)? The return policy is particularly important. It is common practice to over order materials for shutdowns, just in case more are needed. The return policy then takes on increased importance.

If the vendor is new to you, ask for some non-competitive customer references. Ask the customers about delivery, returns experience, stock situation, accounting problems and accuracy. Ask how are the counter/phone people to work with. Is Internet ordering available?

**For large, expensive items more intensive management might be required.**

| | |
|---|---|
| Engineering complete | Physically verify the completion of this step by viewing the drawings, etc. |
| Fabrication has begun | Verify by conversation or by plant visit. Get an estimate of the time in the shop. |
| Fabrication is complete | Verify by visit, inspection of part, etc. Get an estimate of how long it will take for the next step. |
| Testing is complete | How long until it is packed? |
| Packing is complete | How long until it is shipped? |
| Shipment has been made | How long will it take to arrive? Shipment has arrived /cleared customs. |

**Milestone method of managing long lead-time materials** (partially adapted from NSI in resource section)

## Some Special Functions

Questions for selecting a component or other asset rebuilder

- ❑ Is the component updated to the latest engineering revision of the OEM (Original Equipment Manufacturer?
- ❑ Does the OEM certify the rebuilder?
- ❑ Is the component rebuilt to OEM specifications or known to you to be better than OEM?
- ❑ Can the rebuilder give you an analysis of why the component failed, and do they keep records on previous rebuilds of the same component (particularly large components)?
- ❑ Will the rebuilder stock your requirements on a consignment basis?
- ❑ On bulk works, can the rebuilder ramp up to rebuild a mass of items quickly?

**Make versus buy (decision by planner, purchasing, fabrication manager or shutdown manager**

Generally the project manager, in consultation with the maintenance department, makes the decision whether to make or buy a specific part. The decision to make or buy a part depends firstly on the cost comparison and secondly on the availability of the part. Some parts have to be made because the OEM is out of business or the lead-time is both long and unreliable. There are two make versus buy options. One is to find a shop that has the capability to build the part (or even, if you are lucky, builds that item for OEMs as a contractor) or to build the part in-house yourself.

Questions in the make-buy decision include:

- ❑ Which method is cheaper?
- ❑ How long will it take by either method?
- ❑ Do we have the capacity to engineer the part and assure engineering compliance?
- ❑ Will the manufacture or engineering of the part displace other more important jobs (if we do it in-house)?
- ❑ Is there a liability issue that you do not want to take on?
- ❑ Can a core be found on the used market so that just rebuilding will serve instead of building from scratch?

If you decide to make the part, decisions will have to be made:

- ✓ Who will do the engineering?
- ✓ How much time is needed for each step of the process?

- ✓ It there any special liability?
- ✓ Do we use in-house manufacturing or a fabrication vendor?

If we buy the part:

- ✓ Do we go to the OEM (Original Equipment Manufacturer- this is usually the easiest but not always), or to distributors?
- ✓ Should we check used sources, sister plants, or other plants (outside our system)?
- ✓ What is the delivered cost?
- ✓ What is the solid lead-time?

In either event, should the part be stocked or bought specially for the shutdown? If we make them or have them made, can we sell a few to sister plants (or on an EBay auction)? Adding a few pieces to the order dramatically lowers the individual price.

# Chapter 23

# Execution: How to Manage in the Face of Reality

**"Small problems are easy to fix but difficult to see. However, when you let these problems develop, they are easy to see but very difficult to fix."** ***Macchiavelli, Principe (1530)***

## Ready, Get set and Go...

**The final deliverable of all planning is the shutdown package containing:**

1. Shutdown objectives derived from the initial meetings with top management, where goals and objectives are set for the project.
2. Shutdown schedule can be in the form of a Gantt chart showing when activities need to be started and completed. The schedule can be divided up into sections depending on the contract, skill, supervisor, or location.
3. Milestone charts showing where and when in the shutdown outsiders will evaluate the project (one interesting use of milestones is to motivate the team, particularly on longer shutdown events. Letting the team celebrate intermediate milestones might reduce burnout and keep people engaged.)
4. Site plan showing the location of everything.
5. Safety plan with evacuation routes, assembly points, procedures, and locations of safety gear such as first aid stations, emergency showers, and eye wash stations
6. Copies of all jobs that he/she will be supervising with their planned job packages. Between the planned job packages and the turnaround schedule the supervisor can effectively manage their part of the project.

7. Progress report forms that are fed back into the scheduling software —The forms are completed to show where we are in this activity. Some PMS can track partial completion and can check off when resources arrive. On day one, nothing is complete, but materials should be coming in to be staged.
8. Phone facilities, including directories of cell phones, land phones, and beepers. Include all numbers that might be useful (include radio channels). The core of performance is communications. The project office should keep this document current with all contractors and their subs.
9. The work schedule for supervisory personnel so that there is well understood coverage 24/7 for the project.
10. A copy of the turnaround organization is useful to show who is responsible for what. Usually presented in the form of an organizational chart.
11. Instructions for progress reports and meetings. Helps people who have not been through the shutdown before in your facility know what is expected.
12. Meeting or conference call schedule for each team and for general updates.
13. A set of craft work rules or bargaining agreements. It is important to know the answers to questions about turf, helpers, overtime, incidental work, etc.
14. Accident report forms and instructions for how to handle accidents of different sizes and types.
15. Instructions regarding site rules, materials handling, site plans and set down plans for major spares. Within the site plans could be crane lift paths.
16. Instructions regarding the use of contract foreman and clerks
17. Instructions regarding the reporting of contractor daily time reporting
18. A list of helpful reminders like lessons learned from previous shutdowns

## Use of Intranets

Much of the information for the shutdown and most of the shutdown package can and should be published to an internal web site for the shutdown. This requirement is particularly important for information that is expected to change over the course of the event.

The huge advantage of the web site is that everyone can be quickly updated (for example with a new cell phone number or revisions to the plot plan) and everyone on the site has the update. People must be encouraged to use the web site rather than printing out the information. You might also publish a weekly list of items that have been updated.

If an intranet web site is used for managing the mass of paper relating to the shutdown the site should be designed with two aims. The first aim is ease of finding things; the site should be very easy to navigate. During a shutdown people are in a hurry and to get the biggest advantage from the technology they need to get what they want in a few quick clicks. The second is the issue of version control. Each page should give clear information about when it was updated.

The only thing that the web site is not good for is information that should be on paper to gather feedback or that will be handed out such as work orders. Almost everything else is fair game.

## Revisions

It is possible (but usually not desirable) to make revisions to a completed turn around plan. There had better be really good reasons for changes at this point! If changes are made, be sure they are approved at the same level that approved the plan in the first place.

Consider the effects of any changes. If the start date is slipped to accommodate additional production, will storage of incoming materials be a problem? Will anything get rusty, damaged, or offer a theft risk? Study alternatives to accomplish the new objective or accommodate the new constraint without wholesale changes to the plan.

Fight the tendency to overreact to adverse conditions. Be sure to create a trail from the original plan to the new plan and document the change, the justification, and elements of the process.

## Quality

How important is quality for a shutdown? Bad quality has killed people, bad quality has caused forced shutdowns for rework, and bad quality has forced customers to look elsewhere for product. The stakes are high.

Quality has three parts that have to be woven in from shutdown conception to shutdown closure. In this respect quality is like safety, security or cost control. The shutdown plan has to be built on the bedrock of quality (as well as safety, security, cost control etc.) Quality must be written into the mission of the shutdown so that it becomes part of the background for all decisions.

## Quality Planning

Quality has to be built into the shutdown plan. For example, having quality built in might mean requiring proof of proficiency from contract employees before they are allowed on the job site. But it is much more than that.

In regular maintenance work the keys to quality are simple: Right skills, right mental state (not tired, not preoccupied), right supervision, right parts, right tools, proper physical environment (lighting, temperature, etc). These conditions are among the primary reasons for being, of the Maintenance Planning function whether for a shutdown or for conventional maintenance work.

W.E. Deming (1900-1993), the foremost quality expert in the 20th century), said that the greatest motivator for workers was pride in a job well done. The key to quality in shutdown work is to plan and manage work so that the maintenance workers have the opportunity to feel that pride. The shutdown plan must insure that they have the right tools, right parts, etc and are not pulled off jobs, so that they get that feeling of satisfaction from having done a job well. Quality flows from having removed the barriers to quality!

## Quality Measurement

How do you measure quality and who measures quality? In regular maintenance work, quality is measured by the craftspeople themselves (remember pride in a job well done), by their supervisor, and by operations (the customer) when the item is turned over. Quality measurement in a shutdown is very difficult because of the number of work orders and the number of workers. The second issue is that when the job is finished it is hard to see what was done.

Of course, industries where there are significant life safety, environmental or public safety hazards have worked out the quality inspection problem. They build hold points into the job plan where an inspector looks at the work completed before it is covered up, and they have an explicit inspection prior to job acceptance. These plants might also order NDT for welds and other quality issues that cannot easily be seen.

## Hold Points

When a major item is installed, a hold point is written into the schedule to hold for quality inspection. This hold may be required by the OEM (such as a power turbine manufacturer) to initiate the warrantee. The inspection might include vibration readings, amp readings, other NDT. The customer, the OEM, or their representative releases the hold with a sign-off.

Many groups provide support for the quality effort. One of the biggest contributors to quality is the logistics function. Logistics inspects incoming materials. Both bad materials and wrong materials compromise quality. The quality techniques for the shutdown need to be compatible with all the quality efforts that are under way for all maintenance work and to be derived from the quality system for all activity at that facility.

## Quality Control

The last aspect of quality is control. Quality control consists in making random inspections, performing NDT on welds, and keeping supervisors on the job floor. The control aspects again are built into the shutdown plan with scheduled activities like pressure testing, vibration checks, amp readings, etc.

## Vigilance is necessary to maintain quality and safety

Scott Santangelo from Toray Plastics (America), Inc. had an interesting story that should be instructive to all of us. He described what happened in his words. "During a 5-day shutdown, there was a task to replace a Tenter chain as part of a yearly PM. The chain is much like bicycle chain but with links the size of a melon. The total weight of the 2000 links is in excess of 29,000 lbs.

This task involves significant labor, rigging, and coordination. During the planning stages of this work, people were selected and trained to properly check the connections. This work was completed without a hitch. As part of the installation of the rebuilt chain, the links are installed back into the machine in 95 link sections. The crew installing the chain typically marks the connection point between the sections and it is triple checked by other parties to confirm that the connections are properly torqued and assembled correctly. This chain runs at 1000 FPM, so a failure would be catastrophic.

During the last evening of the shutdown, one of the engineers concerned with a section of the track that supports the Tenter chain, which had not been able to look at that section prior to the chain installation. Without any discussion or planning, this engineer, with the help of one of the night maintenance technicians, proceeded to unbolt a section of the Tenter chain to inspect the track. When the work was completed, the two employees thought they had tightened all the bolts and torqued them properly.

That evening, about 1 hour after the production line started up, I received a 911 page (serious catastrophe at the plant). The chain had broken its supporting track and had smashed its way out of the machine. The machine continued to tear through the 250-foot long piece of equipment, destroying everything in its path, until the drive motor finally overloaded.

A veteran production supervisor had been walking past when the catastrophe occurred. The chain had whipped past him within a few feet at full speed. Within 15 minutes I was at the plant trying to speak with the supervisor who was pale in the face and couldn't speak.

With two other engineers, I spent Friday night through Monday morning organizing a clean up, designing the track and machine repair, and replacing the chain with a spare. We had the chain running again by 9 am on Monday morning.

Following the investigation of the accident, the fault was ultimately laid on the engineer who had insisted on breaking the connection on the chain for an inspection. He remembered the chain link number that he had disassembled that Friday night before startup. That information and that link number identified the only link uncoupled and enabled the investigation team to discover how the accident had happened. There was one bolt on the link connection that was improperly torqued and it had come loose while the chain was running at 1000 FPM!

The countermeasure for the future was never to permit emergency, last minute work to occur without key shutdown planning members being involved. First, if this subject had been discussed with the right group, people would have been assigned to complete the proper inspection of the connection on the disassembled link. Second, the inspection process was improved by adding four people to the inspection of every connection made during the installation process. Each person is assigned a different colored paint marker to mark the connection, indicating that it had been inspected.

The inspection data is recorded as part of history in the event that another failure occurs. We felt that this thorough inspection process was necessary to assure the safety of people and equipment."

## Shift schedules

Shift schedules are an important choice. All the factors must be balanced including the cost (overtime and double time ratio in chart below), the fatigue ratio and available hours, the need to get the work done, quality, safety, and the scarcity of skilled labor.

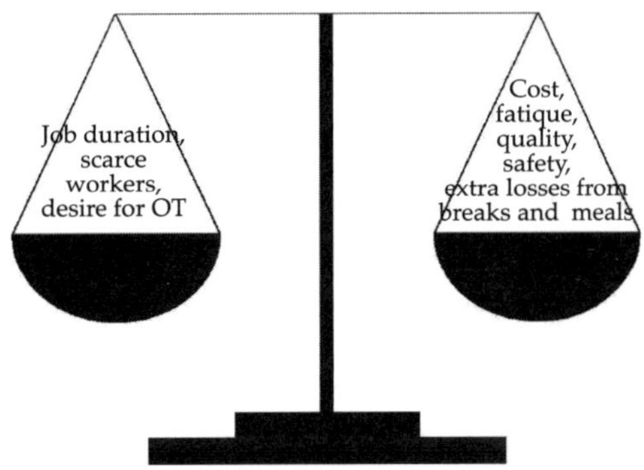


Overtime is one of the big motivators for all maintenance workers and contractor employees (especially if they are away from home). They would be reluctant to take jobs that did not result in good amounts of pay to compensate for the time away from home.

The chart that follows has some rules of thumb and some calculations to show the practical impact of various shift schedules.

The **available ratio** relates the work time after subtracting established meals and breaks. So an 8-hour shift is 1.0 and a 12-hour shift adds 4 hours but also a meal break.

The cost ratio only adds in factors for time and a half and double time where appropriate (USA custom are used and they might be different in other plants or areas).

The **Fatigue ratios** are rules of thumb for people working continuously. In any shift pattern, a two-day weekend rests the employee enough so that he or she can start fresh.

## How to use nighttime

Some schedules use the night shift as the primary time to catch up on critical path activities that have gone behind schedule.

| Hour/Shift | Available Ratio meals/ breaks | Cost for overtime and double time | Fatigue ratio week 2 | Fatigue ratio week 3 | Fatigue ratio week 4 |
|---|---|---|---|---|---|
| 8 Hour 5 Day | 1.0 | 1.00 | 1.0 | 1.0 | 1.0 |
| 8 Hr- 7 Day | 1.0 | 1.21 | 0.90 | 0.80 | 0.73 |
| 10 Hr- 5 Day | 0.95 | 1.16 | 1.0 | 1.0 | 1.0 |
| 10 Hr -7 Day | 0.95 | 1.51 | 0.85 | 0.73 | 0.61 |
| 12 Hr -5 Day | 0.83 | 1.40 | 1.0 | 1.0 | 1.0 |
| 12 Hr -7 Day | 0.83 | 1.60 | 0.80 | 0.64 | 0.51 |

**Shift schedules: An important choice:** (based on work by Mike Brown of NSI)

Nights are sometimes a problem because there is skeleton support as compared with daytime. Interestingly, some gang bosses prefer nights and feel productivity is greater because there are fewer people to interfere with the work (the flip side to no support). Of course, if your facility is near residential populations you might find that there is little incentive to night work because many activities (such as those that produce noise, flash, etc), are prohibited

### The first execution step of any shutdown is to close down the plant!

Closing down the plant seems simple but it is the key that makes the shutdown occur smoothly. An article in the Plant Shutdown Book (by PEM in a source section) lists the important issues of shut-ting down:

1. Notify vendors of gases or other utilities that you are shutting down. Notify processes located elsewhere, who are connected to you by pipeline, that you are shutting down.
2. Decide what is to happen to product in process? Can it be completed?
3. Is it necessary to dump intermediate steps? Are there any hazards?
4. Decontaminate pipe, valves, fittings, and pumps.
5. Isolate plant
6. Run down, bleed, or empty services (compressed air, feed gases, water, capacitors, etc).
7. Isolate systems and verify isolation is effective.
8. Hand over to shutdown team

One way to slip a few extra jobs into a shutdown with a controlled duration is a phased turnover. Create a schedule of exactly how long different processes take to cool down, or get decontaminated. Instead of waiting for everything to be cool, use a phased approach and turn over items as they become safe. You might pick up a few hours to a day of extra work time on those items that cool more quickly. The same phased approach works on startup also. In any event, be sure you are not putting the workers into an unnecessarily risky situation.

### Lessons learned during execution on one shutdown

Larry Jones is a 34-year veteran of Duke Power and has been an outage manager for 8 years. When we spoke he had just come off of an $110,000,000, 12-week utility outage. The planning for that event took 36 months. The cost of having a power plant off line is enormous, so every minute counts.

Larry described the biggest lessons as being concerned with the issue of detailed planning verses loose planning. He learned that the bigger the shutdown the more important was tightness in planning. For example, he planned his set down areas on an hour-by-hour basis because on some days the equipment needing a location for set down changed very rapidly.

Larry's other advice concerned how contractors are used. He learned to lean on his contractors to identify problems and solve them as well. When a problem crops up he expects a contractor to come to a meeting with the solution. To structure this effort, each contractor submitted accurate schedules of all jobs twice weekly. Any problems had to be mentioned and solved. As shutdown manager he verified the status before the meeting so that he knew immediately if the contractor was trying to give him a snow job.

### There are some great tidbits of wisdom from the Project management folks

Some basic Project Management tips from Jeff Crow are useful to maintenance shutdown managers. He says, "monitoring some projects is like herding cats; as soon as you get one piece pointed in the right direction, the others will find some new mischief to get into." In other words, each activity has its own owners, its own priorities, and its own axe to grind (which might not be the same as yours).

Think about a shutdown with 5 or 6 subcontractors. Each of them is looking out for themselves, making sure the business is working (cash flow is positive), watching that the other subs are not taking advantage of them, watching that the prime contractor is also not taking advantage of them, and completing their part of the job.

He goes on, "Still, you do have to monitor your projects in order to manage them. Tracking progress on a project should be a regular part of your daily routine, even if you have other duties that require your attention."

He says, "For some of you the Project will be a part time job to be done along with your many other duties." That might be true during the planning stages but is not often true for shutdowns while they are happening. Whatever the situation, establish a habit of going over the project for at least a half hour every morning.

Track the differences between what was planned and what is actually happening.

This tracking includes:

- Verify that start and finish dates for activities are being met particularly for events on or near the critical path
- Know how cost estimates are working out in reality
- Know whether planned resource requirements are matching actual utilization
- Check whether the expected outputs are being created (are various jobs and smaller projects getting done?).
- In the simplest form, does the crew (either contractor or in-house) show up when scheduled, work the correct duration, and get done when scheduled?

## Daily schedule-- How do you track these differences without a daily schedule?

The daily schedule is the key to success in running a shutdown. The daily schedule is generated from the master schedule, and is designed in two parts. The first part is the list of jobs that can be run the next day. The second part is the assignment of people or crews to each job. The key is to control the shutdown on a job-by-job basis and intervene, when necessary, soon enough to make a difference in that job.

| Work Order | Asset | Craft | Description | 7am-11am | 11am-3pm | 3pm-7pm | 7pm-11pm | 11pm-7am |
|---|---|---|---|---|---|---|---|---|
| 678 | Kiln | Mason | Re-brick S82 | | | | | |
| 934 | Press 6 | Mech | Rebuild ways & gibs | | | | | |
| P110 | Fill line | Elect | Install new filling line | | | | | |
| P110 | Fill Line | Mason | Complete grout line | | | | | |
| P110 | Fill Line | Mech | Assemble Fill Machine and set | | | | | |
| 333 | Conv | Mech | Repair idler SD-234 | | | | | |

The daily schedule is where all your work comes together. All you're planning up to that point will be proven when the worker, tools, materials, and unit to be serviced, converge when you say they will. The process of assigning resources to critical path items first, (items with low float), applies to the daily schedule. The day shift supervisor can easily see that a problem is developing if by lunchtime the mechanics are not deeply into assembling the Filling machine. S/He might ask questions, add resources, and let the job run late, depending on the whole schedule and the situation that day. What s/he does not want to do is to interrupt any job that has been started unless it is absolutely necessary.

### Emergent and Added work

The shutdown planning team always has imperfect information. The team can't see how many tubes are thin, how many tubes will be destroyed while replacing the bad tubes, or that the seemingly solid pump is ready to give up the ghost. The plan is based on what people knew or guessed at the time the plan was made. Some shutdowns experience an additional of 15% to 40% emergent work.

The other issue is that shutdowns are magnets for people's pet projects, deferred items, forgotten items, wish lists, bosses whims, and other good, bad, or indifferent jobs. The cut-off might be 90 days before the shutdown but work (that must be done) trickles in right up until the shutdown is completed! In truth, some of these jobs really do have to be done.

Both of these categories have to be strictly managed or they will crash all your hard work in planning and scheduling the event. The process of adding jobs into the approved job list must have high-level approval, budget augmentation, and possible duration increases.

These requests need to be logged and accounted for outside the base plans so that their impact can be evaluated.

Control: to exercise restraining or directing influence over, to have power over (Webster). We exercise control in a project to force the teams to execute the plans according to the timetable in the schedule. Control is a discipline to insure that what was intended is done.

Control is the discipline that delivers projects on time and within budget. If the underpinnings of job plans and project schedules have integrity, then control will deliver the project on time and within budget.

One excellent application of control is embodied by the habits of the shutdown manager. These habits put the manager in personal contact with all the key players every day. Although discussion is important (and will be considered again later) just being there is half the battle.

## Manager's daily tour routine establishes good practices

1. Visit key people daily. Ask them first, what has gone wrong in the items scheduled since the last meeting, and what problems are anticipated to arise in the near future. It is essential that the key people are free to state what went wrong and what they plan to do about it.
2. Ask what has happened since yesterday?
3. Check in with your point people who are responsible for safety, logistics, engineering, operations (permitting), and quality, for issues and discrepancies.
4. Look at the daily work program, and what is supposed to happen today?
5. The daily project update and control meeting should update the master schedule with everything that has happened since the last meeting. Check that the update covers all activities. Estimate the percent duration left on all jobs that have been started and are on or near the critical path.
6. Supply a short update to the plant manager.

When we each think of control we think of different things. In shutdowns, control means looking at the specifics that, if known, will make a difference in the outcome. Some suggestions to this end are adapted from Managing Projects- (resource section)

**Detailed actions to be a "thermostat" and control the shutdown**

A. **Status (measure the temperature)**
- Determine what the plan says should be happening
- Find out what is actually happening
- Estimate the deviation between the plan and reality

B. **Analysis (is it too hot)**
- Evaluate the impact of the deviation on future activities
- Decide if the impact is a problem
- Determine the cause of the deviation

C. **Management (turn on the blower or open the vent)**
- Find out who is accountable (not who is to blame –very different)
- Decide what action is indicated
- Develop and analyze the solutions
- Take action, or if you need approval, elevate the problem with your own analysis of the potential solutions
- Implement the solution
- Follow up to see if the solution worked
- Go back to checking status

**Control Metrics** (partially adapted from Project Management Mentors).

How do we know if we are in control of a shutdown? Are there metrics (measurements) that will allow us to compare this shutdown with a prior one? Can these metrics be used to compare several plants in the same company?

Charts could be kept in the shutdown office for some of the items mentioned below. Usually only a few good benchmarks that are meaningful to your shutdown should be measured and discussed at shutdown meetings

**Examples of Exception data (during the execution phase)**
- Critical path activity missing deadline (how much, how many, and the discussion would be about what is being done to put the shutdown back on track)

- Activity slipping LST (this will force the path into the critical category unless the duration of the activity is accelerated)
- Activity slipping LFT (will impact the critical path and completion. Review the next activities to see if any of them can be accelerated)
- Any other activity on critical path being delayed
- Activity where deadlines have been revised numerous times (indicating that a problem is brewing. Intensive management might be necessary to keep this activity on track)
- Added work
- Emergent work

### Staffing

- More staff being used than were budgeted. (If activities are getting done on an accelerated basis, extra staff is not necessarily a problem. It is a problem if the activities are also behind or even only on time)

- Staff not available to fill out budgeted amount. If the staff shortage is in a critical area then a problem is brewing and more emphasis should be placed on the staffing issue. Additional overtime might be an option.

- Staff is being pulled off for other work. This condition indicates a disturbance in the communications, or other defects in leadership of the plant or of the shutdown. The policy team or the plant manager might have to intervene).

- Lack of specific skill sets. As mentioned above, this problem could be extremely critical and might require a creative solution (like flying in a crew from another country)

### Budget

- Spending too much (compared to date in shutdown spending schedule)

- Spending too little (sometimes can be a problem depending on why)

- Large costs not in budget at all. A serious problem with the original estimate that might require an emergency meeting with the policy team to beg for more money.)

### Close the feedback loop

How many of us design a beautiful project plan and don't expend the slightest effort to close the loop with feedback? All the feedback will no doubt disrupt the perfection that we created and make the plan messy and out of balance.

Mr. Crow admonishes project leaders to do their job. "You, as the project leader, have the responsibility of tracking the project. If you are not receiving the information you need, you must go and get it. If you set an expectation, for example, that status reports are to be submitted weekly, you must follow-up on it."

The toughest issue is that project managers tend to monitor most closely the activities within their expertise. There is a pitfall here. "Monitoring the technical aspects of a project is usually where the energy is focused. However, if all the attention is placed on technical measurement there is a strong likelihood that the things that will cause problems in the project will be team and interpersonal issues. Project monitoring is a process. It needs to be done constantly and consistently."

Control is the sum total of all the things a manager does to increase the probability that the planned results of the organization will be achieved.

### Peter Principle can strike the best-planned Shutdown

The Peter Principle says that people will rise until they reach their level of incompetence. As people take on assignments they will eventually face an assignment that is bigger than they are. Preben Ormen puts it succinctly "What comes easy to some is hard for others. Sooner or later everyone on the team will run into something that stumps him or her.

Often when this happens, the task will be put away somewhere while the team member thinks about what to do. Sometimes nothing comes to mind or the team member simply gives up, especially if burn out is setting in.

Such tendencies must be caught early and no one must be allowed to get away with it. You need to make sure people feel comfortable bringing their problems forward and actively asking for help if they get stuck. You must get all problems out in the open, use the issue log to keep track of them, and get your people the help they need to get the job done."

### Feedback is the breakfast of Champions!

The key to tracking events is to manage the information coming in about the status of every job. It is essential to feed back the job completions and job problems to the PMS or CMMS (if they are linked) software, at a frequency that will make a difference to the duration of a shutdown.

### How often to check the status of all jobs running and update the schedule

| Duration of shutdown | Reporting interval |
|---|---|
| 36-hour and shorter shutdown | 2-hours |
| 7-day shutdown | 8-hours (1 shift) |
| 15 day and longer | 24-hour (daily) |

Keep your ears open because the reason for not being careful with the update is always the same. You'll be told that your team doesn't have time to check the jobs that often. Like many things, doing it right the first time takes less time than doing it over later. It takes less time to make an inquiry about a job status on every shift than it will to fix a major problem that is threatening to delay the shutdown, two days after the fact. Ultimately it is the shutdown manager's responsibility to make sure good information is secured, on which to make the best decision.

In a large shutdown the management crew might need someone whose primary job is to keep the schedule accurate. New schedules might be printed once a day or once a week, but the software is 100% up to date on the schedule above. The question is, in what ways can we achieve the schedule update without excessive effort?

### Techniques to gather the information to update the schedule:

- Post a schedule on the wall and have the 'responsible parties' update it in colored ink before every morning meeting.
- Organize a formal update conference call at an agreed time; review all jobs running with a quick code (such as done means

accepted by quality, completed means work is completed but not accepted, 50% means an estimate of half done etc.) You must suppress the tendency for people to tell stories about why whatever is happening or you'll be on the call all day.

- Have a planner, scheduler or clerk call every 'responsible party,' find out status, and enter the data and update the schedule (before the meeting). This check should include a call to logistics on critical materials and rental equipment not yet under their control. Critical path jobs should be physically verified. As a famous diplomat said (in a different context) "trust but verify."
- Set-up an E-mail account for schedule updates. Everyone should be told to E-mail job status reports by 8 AM.
- Be rigorous about closing work orders in your CMMS (Computerized Maintenance Management System), which will update the PMS.
- Collect status reports during your morning tour and update PMS at the end of the tour.

**What information needs to be updated?**

Remember, you must collect details of individual jobs (or even individual activities), individual parts, or individual machines to manage the problems that crop up. In other words, problems crop up in the details and have to be dealt with in the details. Questions to ask are:

- Is job xxx done and accepted?
- Has xxx material come in yet?
- Are you ahead or behind schedule for that moment?
- Are you over or under budget?
- Are the jobs running as estimated?
- Are any clouds on the horizon or problems brewing?

The shutdown situation is reviewed periodically. Well before the shutdown starts, this meeting could be monthly. The function is to get everyone on the same page and hear what is going on from each person. As the shutdown gets closer, the meetings might jump to weekly. A complete shutdown review meeting might be held weekly (for a several week shutdown), or more often for a shorter event.

Once the event begins, this meeting can take place daily in abbreviated form as a status meeting. The other common alternative is to supplement the formal meeting described below with shorter, more detailed, daily sessions. In these sessions a smaller group hashes through the details of any specific problems. One common trap is that the details are gone through at the Shutdown Review Meeting. This arrangement makes the meeting very long, and for people not directly involved with the problem, very boring.

### Shutdown Review Meeting Agenda

1. Review of the mission statement
2. Scorecard: Brief review of where we are today in relation to schedule
3. Scorecard: Review of status to budget
4. Review of action items and promises made at the last Shutdown Review Meeting
5. 5 minute (or less) reports from job or area managers
6. Accomplishments and acknowledgements
7. New business (problems or opportunities not presented before)
8. Detailed view of problems and mitigation, including requests for resources or other attention
9. Quick overview of major breakdowns brewing and hot upcoming topics
10. Plans for next period

### Private discussion guide

Some people like to give bad news in private. They might know of a problem but not want to share it at a meeting. They may become aware of a problem in another area and not want to embarrass the area manager publicly. Also there is information that is not necessarily relevant to the larger group, but is very relevant to the shutdown manager. The Project Management Mentors recommend a structured approach to a Q & A with the accountability holders with thought-provoking questions. This conversation would take place privately, before the status meeting, so that subjects that are uncovered can be aired and discussed (if appropriate), such as:

- Do you foresee any problems coming down the pike?
- Is there anything I should know privately about the shutdown that should not be discussed in the review meeting directly but should be discussed in the background?
- Are resources being moved away to do other jobs?
- Will those moves impact your jobs (particularly ones with minimal or no float)?
- Is your team working productively?
- Is the team 'happy,' and do they need anything?
- Is anything missing that would make your job easier if supplied?
- Is anything bothering you?
- Have you or your team had any ideas that might help other teams?

## When good shutdowns go bad

From observation of shutdowns you can see that the ones that get into trouble have certain things in common. A bad plan or a monstrous emergent problem, not on anyone's radar, can sink an otherwise decent shutdown team. Also, policy teams that cannot say no to added work can deep six a perfectly good shutdown plan. But the most painful problem is with faults in the leadership of the Shutdown manager. So we call this problem:

**Pitfalls in leadership** (partially adapted from Managing Projects by John Reddish to suit shutdowns). These pitfalls occur when you, the leader:

- Try to do everything yourself
- Keep up a running battle with your team
- Keep up a running battle with your boss
- Keep up a running battle with operations (or your customer)
- Keep up a running battle with the contractor
- Keep busy with details
- Fail to attend shutdown meetings
- Fail to visit the field
- Fail to maintain the schedule and budget updates
- Have really long meetings and go over things over and over again
- Do not follow-up on decisions and promises from the last meeting

- Avoid make decisions
- Ignore ideas and feelings of others
- Keep information to yourself
- Share too much detail
- Emphasize production, not people
- Take credit for the good work of the team
- Surprise your customer with your results

Problems are common with shutdowns. The seeds of these problems can sometimes be seen in the planning stage if you are looking for them. In the same way that an acorn is very different from an oak tree, the seed of the problem might be very different of how it will turn out. These problems are not visible during the planning phase as problems unless you are looking for the seeds and you are alert.

### Seeds are sown, and if the problems grow they will show up during execution

- Not keeping the plan up-to-date, or starting with a poorly defined shutdown causes the shutdown plan to be seen as just a chart on the wall, not a living, breathing, structure. The "chart on the wall" attitude results in people going through the motions of management (because they have a lack of buy-in to the plan). Of course people don't manage their areas if they have:

- Inability to stop scope creep. Did you ever try to satisfy kids by giving in and getting them what they've been whining about? This response has about the same medium term effect of giving in to people who want to add work to a shutdown. They are satisfied for about 30 minutes. The failure to say no is sometimes the result of inadequate leadership from the Policy Team. Linked to this oversight is a failure to manage everyone's expectations in line with what it is possible to get done.

- Inadequate planning with missing critical steps, or failure to identify necessary tasks (insufficient task breakdown). These lacks may be the result of a planning cycle that is too short, lack of knowledge in planning, lack of relevant job knowledge, or lack of expectations by the leadership.

Some of the symptoms include failure to anticipate collisions between space, labor, or equipment.

- Unrealistic targets or insufficient resources dedicated to the project (at any level) to make those targets. The policy team might have their heads in the clouds or be remembering when labor cost only $10 an hour. Related to this shortsightedness are unrealistic time frames with a contractor who tells you what you want to hear, while they know the truth. For instance, a contractor might try to take advantage of the situation with a low-ball estimate.

- An inflexible plan that is unable to assimilate new information. This situation may be comparable with someone who gets in over his or her head and doesn't ask for help (or doesn't even know there is a problem)!

**The last step of every shutdown is starting up.**

The issues of plant start up after shutdowns are well known. Pablo Cussatti, of Pepsi Bottling Group reported that, in the past, "the general rule when I got to the plant was that after a shutdown it would take two weeks to get back up to previous efficiencies." Two weeks of lost or sub-quality production at a Pepsi bottling plant represents a good deal of money.

Mr. Cussatti and the start up team worked to solve the problem. He goes on to say, "We created a shutdown schedule that was reviewed daily; we also created a start-up checklist to focus on the areas that were worked on during the shutdown. We scheduled re-work product to be run on the equipment that was worked on, to assure that actual production runs would go smoothly."

Mr. Cussatti's team was careful to avoid iatrogenic failures (failures caused by the service person) by working on only the items addressed during the shutdown. "The audit portion is necessary to ensure that only parts and areas that require work are worked on. This approach eliminates unnecessary repairs that could cause more harm to the equipment. A start-up check list is also recommended to avoid the 2-week start-up curve."

**The Plant Shutdown Handbook (PEM in Resource Section) ists a few of the issues of start-up.** The important issue is that, like the shutdown list, the start-up list must be tailored to suit each plant. It is necessary to:

1. De-isolate systems
2. Clear all permits
3. Return services to on-line status. Remove all blinds (arrange for positive verification of each blind removal).
4. Run plant function tests.
5. Create a punch list of problems indicated by function tests, inspections, and incomplete work.
6. Remedy punch list problems that will have an impact on startup.
7. Re-commission the plant
8. Re-introduce inventory to system in a structured way.
9. Verify product output quality, run up to full output
10. Resume stabilized full production

CHAPTER

# 24

# MASTER SCHEDULE FOR PHASE 3 EXECUTION

## EXECUTION OF THE SHUTDOWN, OUTAGE, OR TURNAROUND

### PHASE 3

Production and execution process includes building, assembly, removal, and replacement. The phase ends with substantial completion of shutdown. Delivery of punch list is the last event of phase 3.

## SHUTDOWN MONITORING SYSTEM

Whatever strategy is used, you need to be able to anticipate problems before they impact the critical path. The monitoring system has to be detailed enough that you can, if there is a problem, be made aware of the problem and able to intervene in time to solve it (and not hear about it a week later).

| Do I need this? | Item |
|---|---|
| | 1. The logistics function should be in control of the site. Use the plot plan to stage large parts. Manage all rental equipment, crew quarters, and site security. Parts are starting to flow into the logistics function, in some instances by the hundreds or thousands. Verify that the system set up is adequate for the job. Sometimes the logistics function is part of the main contractor's duties. |
| | 2. Put barricades into place. They should be designed to provide safe pathways through construction areas, restrict contractors to areas, and limit access to dangerous and sensitive areas. From time to time, inspect barricades to make sure they are still in place. |
| | 3. Put up signage to direct the contractor to parking areas, work areas, relief areas, and assembly points. Signage for trucks to receiving is also needed. Be sure security is inplace. |

| | |
|---|---|
| | 4. If possible, erect all scaffolding before work begins. If several vendors are used, require that they clearly mark each piece of their scaffolding (even better, Mike Brown recommends assigning colors to each scaffold vendor and having them spray their units that color). |
| | 5. Perform quality inspections on all scaffolding to insure safety rules have been followed. Repeat these inspections whenever new scaffolding is erected. |
| | 6. Deal with traffic control including incoming heavy trucks, very heavy and large loads, and any excessive traffic your shutdown puts on local streets. Hire flag people or (in some areas) contract with local police for traffic control on public streets during peak times. |
| | 7. Verify all safety items are in place, including extra fire fighting gear, eye wash stations, retrieval gear, decontamination rooms, first aid stations, emergency equipment, rescue kits, etc. |
| | 8. If contractor safety supervisors have been agreed upon, they should schedule a meeting with site safety people and review (or negotiate) shared responsibilities. |
| | 9. Logistics. A contractor has to manage sending bulk works (including wholesale rebuilding, testing, certification of pressure relief valves, pumps, valves, motors, steam traps, bursting disks, etc.) to vendors and manage receiving of completed items coming back. Be sure to distinguish between the outgoing and the incoming. |
| | 10. Provide a cell phone, landline phone number, and beeper list of personnel and contractors for the shutdown, along with other frequently used numbers (fire, hospital, etc.). Include numbers for other plant services that might be useful such as quality control, security, employee health, calibration, etc. Include radio bands for safety, security, maintenance, operations, and others. |
| | 11. Secure a list of contract workers including total numbers for each phase or day, and their names. Each contractor employee should be required to sign in and out each day. It is desirable, for an accurate detailed history, that each contractor or his or her supervisor will also sign on and off every work order. |
| | 12. Start to conduct random drug and alcohol tests per agreement in your contract. Continue throughout the shutdown. Repeat tests after anyone returns from leave. |
| | 13. Compare the appearance of the people who show up with the people on your approved list (that's why you took the pictures). Arrange a charge back for employee substitution unless you were informed. If you have any questions about someone's competence, give him or her a skills test right then, before they have an opportunity to ruin any equipment |
| | 14. Determine the number of forklift(s) and mobile cranes, as well as any other mobile equipment, in the proper capacity. Have them parked where indicated on the site plan. Inspect, fuel, and verify daily. |
| | 15. Be sure all safety training is complete before people are allowed on site. Be sure everyone necessaryhas fork truck training. |

| | |
|---|---|
| | 16. The Turnaround Package should contain<br>• Turnaround objectives and mission statement from policy team<br>• Copies of Gantt charts showing the activities taking place until after the next meeting (or the whole shutdown if it is simple enough to understand)<br>• Copies of all work orders that he/she will be supervising with planning packages as appropriate<br>• Progress report forms—completed to the point applicable<br>• Work lists for areas nearby<br>• Site drawing showing where everything is kept including cranes, large parts, portable bathrooms, assembly points, etc<br>• Telephones, radio channels and beeper list<br>• The work schedule for supervisory personnel<br>• A copy of the turnaround organization, showing a clear chain of command<br>• Instructions for progress reports and meetings<br>• A set of craft work rules<br>• Instructions regarding materials handling including summary of heavy lifts and lift paths<br>• Instructions regarding the use of contract foreman and clerks<br>• Instructions regarding accidents, damage, and spill reporting<br>• Instructions about proper removal and disposal of hazardous materials. Keep in mind the issues of chain of custody (you are the responsible party forever so you had better know it was disposed of properly)<br>• Instructions regarding the reporting of contractor daily time reporting<br>• Site rules and regulations<br>• A list of helpful reminders and lessons learned<br>• Summary of cost account codes for the turnaround<br>• Hazard Safety analysis including PPE requirements, fall protection, attachment points, and safety contact people.<br>• Preliminary blind list (use a two signature system- one to add a blind and a second signature when that blind is removed) Be sure every individual blind is explicitly identified. |
| | 17. Prepare the building to protect other processes from dirt, sound, debris, chemicals, etc such as putting up temporary walls, plastics, and caution tape. Be sure to include dust, flash, and noise control for neighbors where possible. |
| | 18. Convene an all-supervisor, all-engineer, all-operations etc. meeting and review the overall job plan. Be specific about the opening salvo (what will happen first, and who is responsible). Introduce KPIs for this shutdown along with the goals in the plan. |
| | 19. Clean, cool down, ventilate, wash, and decontaminate areas to be worked on. Test for safe working conditions before issuing initial permits by operations. |
| | 20. As clearances are issued ensure that locks are placed and gang locks are used if appropriate. |
| | 21. Place blinds and use a two-signature system. Identify each blind. Some firms spray paint the flange where the blind is located to make it easier to find for removal. |

| | |
|---|---|
| | 22. Provide construction protection for machines in process such as bearings, insides of machines or components, open flange faces, transducers, steel items that might rust, etc. |
| | 23. Consider taking archival pictures throughout the shutdown. It also may be useful to show the locations of buried pipes and concealed wiring. Be sure to photograph all milestone events or completions for the shutdown report. |
| | 24. Continuously monitor and post progress (to your software, if you are using any, or to your hand drawn charts) on an activity-by-activity basis. This procedure enables the software to provide real-time alerts when activities fall behind or expenditures exceed budget. Identify slippage promptly and take corrective action. Expedite as necessary. Add resources (overtime or additional contract support) to overcome the shortfalls in order to return to the plan. Focus on and protect the critical path, but keep an eye on the other paths. Make sure they do not exceed their available float. |
| | 25. Keep yourself familiar with lift paths. Design and enforce safe lift paths for all heavy moves. Check that the lift paths and crane placement being used are the same as the plans that were approved. |
| | 26. Scope creep and the resultant Change Orders – should be resisted. 15% - 40% of work in a shutdown emerges after the shutdown has started - be prepared. Some changes won't go away, but the overall project must be protected. Have a MOC (Management of Change) check list. A change order process should be followed from the point that the work list is closed. How many times has it been said, "While we are doing this, it would be easy to also do this other thing"? If change is accepted, everyone up the ladder should sign off on it and the accompanying statement of consequences in terms of extended duration and increased cost. |
| | 27. Manage added work and emergent work into separate accounts so that they can be reported as different line items. These two accounts should be discussed at every policy team meeting. |
| | 28. List all the issues of turning the plant back over to operations. Include resolving all tag-outs, blind lists, open lines, testing issues, etc. This is a primary responsibility of operations with maintenance advising. |
| | 29. Declaration of Project Completion. Start looking for undone items for the punch list |
| | 30. How do you know when you're done? The endpoint was defined during the project planning. . With some projects the endpoint is clear, when the tractor rolls out of the shop or the press starts stamping out parts. Individual maintenance projects have well-defined end points. Collections of maintenance projects (called a shutdown!) might not have such clear endpoints as will a group. |
| | 31. Make sure that complete cleanup of the site is one of the items! Repaint any items needing it. |
| | 32. Take a walk around, Is there any damaged pavement or fencing that needs to be repaired? |
| | 33. Return all materials not used. Return all rented items including lifting equipment, rented tools, welding gas cylinders, trailers, etc. |

# CHAPTER 25

# COMPLETION

Want to get confused? Ask a tradesperson if they have finished a job. Then ask operations if the job is complete. Then ask the planning department if it is closed. The problem is that the meaning of complete is different to different people. When the job is substantially done, the tradesperson considers it complete (there could still be some work left but the machine starts and seems to run). For operations, the job is complete when they lift the clearances and accept the item for production. For the planner, the job is complete when the paperwork is fed back into the system and the work order is closed. In other areas, completion is declared by degree of safety or quality. A completion and approval form may be attached to the job package:

| Work Order Number | Description | | |
|---|---|---|---|
| Job is accepted complete | | | |
| Exceptions | | | |
| Department | Action | Signature | Date |
| Contractor- Mechanical | Job is completed | | |
| Contractor- Electrical | Job is completed | | |
| Safety | Job is safe | | |
| Operations | Job has been accepted. | | |
| | All permits removed. | | |
| | All blinds removed | | |
| Data Entry | Completed to CMMS | | |
| | Completed to PMS | | |
| Comments | | | |

Imagine the complexity of trying to find out what exactly is going on if there are hundreds of jobs running, with hundreds of people involved. Consider some accepted definitions for the stages of completeness (feel free to use your own but just be sure they are well known to all stakeholders in the shutdown including the contractors).

**Done-** The job is completed but there is still some buttoning up and cleaning to do.

**Finished-** The job is mechanically complete, buttoned up and the area is clean.

**Accepted-** The job is finished and the clearances have been lifted, the item is ready to go.

The completion of a major shutdown is really a three-step process.

**Step 1:** The first step of job completion is at the end of phase three. There is a time when the area bosses declare the jobs in their area substantially complete (or are not going to get completed in this shutdown). They start to create lists to be completed. In construction these lists are called punch lists. They are made up of all the little things that need to be done or have been left undone.

This preparation for completion also ends phase three and begins phase four. There are hundreds or thousands of details that have to be managed, and completed. Safety issues dominate at this time.

This time is also dangerous because pressure, steam, heat, and chemicals are reintroduced to the systems. Is the plant safe to operate with reduced safety vigilance (when compared to the vigilance needed during the shutdown)? Safety vigilance will still need to be heightened while the shutdown is being cleaned up and barricades are removed in the next step.

**Step 2:** is completed when the punch list work is finished and the customer is satisfied. This is the core of phase four and could also be called the clean-up phase. This phase might last only a few days or a few weeks at the most (Phase 4 lasts a year on an Aircraft Carrier). One of the most important things is complete cleanup of all shutdown debris, wastes, and the disassembly of scaffolding, barricades, and fencing, and return of all rentals. All unused materials are returned to the stock room, then returned to the vendor if possible (or if desirable). Full, standard, safety practices are resumed on the site.

In a successful plant start-up- although the paperwork and contracts are very important the shutdown is not complete until good product is flowing again. Full, high quality production is one of the three ways a shutdown is closed out. In short, the plant is back to profitable and safe operation.

**Step 3: Internal closeout or Administrative Closeout-** This closeout includes collection of all bills, time information, final drawings, etc. All raw data from the shutdown is collected. Once all the time sheets, contractor bills, and vendor bills are in, a final shutdown budget is prepared. The packaging, indexing, and filing of information is the concern of phase five, which encompasses the completion of paperwork and the preparation of a historical document.

**External closeout or Contract closeout-** only when all work is either done or declared that it will not be done can the contractor submit all the final bills. The initial contract should have required that all bills be submitted within 4 weeks of the completion of the work, ensuring timely submission of all external paperwork.

As part of the external closeout process, a team member has to read the contract and verify that all the work contracted for is completed, or in some way accounted for. When all aspects are complete the contract can be settled.

In any construction project there might be a retainage held for some agreed amount of time. In other instances, an amount might be held in lieu of a warrantee. The third step is also resolution of all outstanding issues with the contractors, including claims, change orders, mistakes, and payment of bills within the terms of the contract.

The third step of completion is also the compilation of all information, bills, closed out contracts, reports, and finalization of spreadsheets, into the shutdown narrative book.

- The shutdown narrative book is intended to tell what happened, and can be many volumes. .
- A post mortem is conducted 3 to 6 months later. It also includes the results of a follow-up meeting, when the team has had a chance to move some distance from the event and gain some perspective.

# CHAPTER 26

# REPORTING

The reporting system has two distinct functions. Its first and foremost function is to keep everyone informed about the status of the shutdown. The second function is to prepare documents that become part of the history of the facility and inform later people about what happened and how it happened.

Most PMS have extensive status reporting and graphical capabilities. Some software was purchased to make 'pretty charts and displays' that management was clamoring for! The ability to help organize and manage the shutdown was a secondary byproduct of the ability to satisfy management's need for charts.

## What is happening?

The need to know is universal. But different people need to know different things. In one older project management system, the supervisor got hundreds of pages of printouts every week. He actually needed and referred to only the first two sheets and threw the rest away.

Project Management Mentors has an interesting generic grid that sheds some light on this issue:

| Report parameters | Top Management | Immediate Manager | Team Members |
|---|---|---|---|
| Levels of Detail | Less detail, more graphic, information tool | Intermediate | Greater detail lists Action tool |
| Timing | Less frequently (Minimum monthly) | Intermediate | More frequently (Minimum weekly) |
| Content | Just the overview Problem isolation and Recommendations | What they request to see | Overview and sections that impact it |

There are three kinds of reports (or three sections of reports):

- Status Reporting: Where we are today

- Progress Reporting: What we have accomplished
- Forecasting: What is likely to happen?

Many reports to management have elements of all three.

### Keys to effective reporting

- KISS (Keep It Simple Silly)
- Audience specific. Make sure the reporting covers the audience's interests in the shutdown.
- Be concise and terse (short) so that people will read the report and not be overwhelmed.
- Use graphics effectively but honestly and do not overdo them.
- Keep copies of reports in project history binder (also called the project narritive).

The issue is called developing a reporting hierarchy. The team has to decide what is the minimum amount of reporting that will satisfy their information needs. Other stakeholders need information updates appropriate to their stake in the outcome.

### Phase 5

This is the phase that helps create a project history in a form that can be passed on to new project teams. The process will help train project experts for all kinds of future projects. The core activities of this phase are to organize existing files, throw away useless files, establish TA dates on certain files (TA or TAD = Throw away dates) and write a lessons learned document. The goal is to promote the creation of a body of knowledge so that future shutdowns are not as costly in either money or downtime, or both.

The worst (and most common situation) is that the project is reduced to boxes of files stashed in a document warehouse (somewhere). The knowledge so expensively learned is lost to the next shutdown. One of the most important clean up tasks is to create an index of all the materials in every box. This index is published along with the shutdown narrative.

On the other hand, in some organizations shutdowns are so much a part of normal business that the body of knowledge of good practices is widely distributed in both the maintenance and user organizations. Their lessons are really learned and serve to inform all projects the organization undertakes.

## Papers and computer files

Some time must be spent organizing the papers from the shutdown. Get advice from your firm's legal council about retention periods for project paperwork (some items might have to be kept for 2, 3, 7 or 30 years or even forever for legal reasons.) Discard everything that you can, but keep the important documents.

Create an index of the papers and computer files from the shutdown. Some shutdowns don't even have effective indexes to find individual papers for review or learning. Again, this process can be ongoing from the conception through the execution of the shutdown. Most files are on computer and can be kept there.

## Files to review and keep:

- Keep the scope documents and estimated budgets generated by top management
- The list of jobs is important because it will save time when you do another shutdown in the same area (this list is input into the next shutdown and is a time saver).
- Keep individual job plans (which can be part of the maintenance planner's library), particularly if there is any possibility that the job (in any form) will be repeated.
- Keep insurance documents and proof of insurance from all contractors.
- Keep closed work orders listing materials used, labor expended, other resources used, and closing comments, which should be entered into CMMS and become part of the unit histories. If notes are added and sketches are scanned, the paper document can be discarded (if allowed by legal advisor).
- Keep contract documents, purchase orders contractor lists, and any evaluations done (can go to purchasing or wherever contractors are hired from)

Schedules, schedule compliance reports, and minutes from compliance meetings can be put into a book form to make them easy to review. Many files are on the computer. Burn CDs for copies.

There are two documents that might be created from a shutdown to preserve the lessons learned. Details of each follow.

## Narrative

One document is the shutdown narrative. Additions are made to this document during the shutdown from the initiation phase and it is completed at about the same time that the shutdown leaves phase four. The narrative is filled with answers to the question: what happened in this shutdown? The document is specific to a specific shutdown, and reading it should enable one to reconstruct what happened. It could be considered a diary for the shutdown. In some organizations the manager of each function keeps a diary of the shutdown from their point of view and these documents become part of the narrative.

The narrative could be used as a teaching tool and almost a classroom guide for new shutdown team members, and of course as a guide for the next time a shutdown is scheduled in that area. Some editing will be required to create a readable narrative and to explain what happened and why. This report is very specific to the specific shutdown, but does not look at some of the bigger issues raised by the event. The shutdown is reviewed in a Post Mortem, which deals with some of the larger issues of the shutdown, and shutdowns in general.

When the narrative is completed, the KPIs for the shutdown are calculated and reported upon. They should occupy a leading position in the narrative.

The time to start building the shutdown narrative document is when the shutdown is conceived. One organization opens a folder on a shared network drive specifically for lessons learned. Everyone is expected to contribute information about the problems and their solutions to this folder. After the shutdown, these documents are edited and compiled. Assignments at the shutdown review meeting include writing up issues for this document to insure that people contribute to the shutdown history. One rule is that no one is allowed to whine or complain without being constructive; they must state the problem and describe the solution that was implemented. They can also report how the solution worked, what was missing, and what will improve the solution for the future.

The narrative would start with the top management mandate for the shutdown, and include the justification, the constraints, and any related documents from phase one. From phase two would be filed documents covering the validated scope of work, all the planning documents (or references to them if they are stored in a stable location not likely to change or go away). Filed or referenced schedules from various periods, the shutdown package items, logistics plans, site plans, material lists, and relevant communications, meeting minutes and progress reports, would also be included.

The filing would include work-arounds for problems, risks and actions, and any safety incidents. This document should be printed, but also kept in a computer readable format because elements of it might be copied for future events. Whole elements can be cut and pasted into the next shutdown structure.

**Lessons learned must be preserved in a Post Mortem.**

The second document is the Lessons Learned report, which is completed 3 to 6 months after the shutdown. In this report, the whole shutdown is discussed in a structured way and any conclusions are recorded. The document answers the question; how could we have done better in this shutdown and what policies and procedures can we put in place to do better in future shutdowns?

Many organizations spend millions on shutdowns and not a penny on learning from them. Phase Five is optional. Remember what Winston Churchill said, (although not in reference to shutdowns) "Those who do not study history are doomed to repeat it."

**Three to six months later…**

Try to get the project team together for a day or two (or at least the key players from both the project side and the user side). To prepare for this meeting, the team members involved should review the shutdown narrative document. They should also read any lessons learned from prior shutdowns to see whether recommended improvements were made, and if not, why not.

Meeting minutes from the whole shutdown process from inception to closeout should be reviewed, and one person or a small team should be assigned to write what lessons were learned, with references to the minutes from the status meetings and with to source documents where appropriate.

This is a time when the team has should be rested and reached some perspective concerning events in the shutdown. Some fundamental issues are to be discussed here and it is essential that it not become a witch hunt or a blame game but an authentic inquiry into what really happened, what processes worked and didn't work, and how could improvements be made in the future.

At an initial meeting, discuss generally what worked and what did not work. The report should start with an executive summary, with the high points limited to one page.

**In this meeting, discuss and make notes on:**

- Did the shutdown policies impact the shutdown for better or worse?
- In what way did the shutdown meet or not meet the customer's needs and expectations (and management's expectations) Be specific?
- Was engineering adequate for the project?
- Were planning and scheduling adequate for the size, complexity, and repetitive nature of the shutdown?
- How much did the shutdown change between the time it started and when it finished?
- How close were the budget and the baseline schedule to the final results?
- What worked and what didn't work about the communications between shutdown team members, between the team and the customer, and between the team and the contractors
- How did shutdown and start-up of the plant go?
- What did and what did not work in the shutdown team allocation of responsibilities. On the same subject, were all parts of the shutdown responsibilities handled?
- What techniques or technologies helped this shutdown (and which ones should be dropped or what should be added in the future)

**Lessons learned may also be in the following specific areas:**

- Shutdown policies, overall evaluation of policy team
- Work scope, added work, emergent work
- Team work (in and between the major teams- planning logistics, engineering, execution, etc)
- Job planning (in general)
- Shutdown planning (and coordination of jobs)
- Cost performance (how good were estimates compared with the actuals?)
- Shifts and work schedule
- Work completed compared to work scheduled
- Scheduling accuracy, completeness, compliance
- Safety and environmental compliance
- Contractor performance, performance of bidding system
    - Vendor performance, purchasing system
    - Quality, measurement, control, management

Keep meeting minutes from this whole process and series of meetings. Record what lessons were learned with references to the minutes from the post shutdown meetings. Many great ideas will occur and be discussed at these meetings and it pays to preserve them.

## Chapter Headings

In the discussions, the lessons learned may come out in several areas. Some suggested headings for these lessons (do not limit yourself to these items) follow the phases of the shutdown itself. Specific areas or chapter headings in the final document might be:

- Executive summary
- Shutdown metrics including duration, performance to budget, jobs complete percentages, safety, and start-up.
- Initiation phase issues
- Planning phase issues. How many hours did it take to plan this shutdown
- Engineering Issues
- Execution phase issues with a focus on:
    - ♦ Safety including statistics on accidents, near misses, and issues with permits, PPE. Include detailed summaries of serious accidents as appendices.
    - ♦ Environmental incidents include summaries for any serious in-cidents in the appendix.
    - ♦ Logistics Issues
    - ♦ Contractor and vendor performance
    - ♦ Added work and emergent work
    - ♦ Schedule compliance including schedule breaks, logistics issues, mobile equipment issues. Any specific items that caused delays
    - ♦ Execution and work assignments including a review of assignments, resource issues and execution issues
    - ♦ Close out or completion phase issues. How did the start-up go?

CHAPTER

27

# MASTER SCHEDULE FOR PHASE 4 COMPLETION AND FOR PHASE 5 CLOSE OUT

| |
|---|
| **Shutdown accepted by end-user, start-up completed**<br>**Phase 4** |
| Turn-over, quality assurance, life safety testing, start-up, punch list completion, restore to safe, quality operation |
| **Define how you will know when a shutdown is completed.** |
| Go back to the Policy Team's shutdown objective. Could this goal be the endpoint? Is the promised startup date an end point? Is running out of money an endpoint? Early on, this point should be discussed and decided upon. |
| **Think about the next one** |
| Start thinking about and collecting information for the next shutdown. There is always more to do. Collect any logs and notes from individuals; be sure intermediate papers are not discarded before they are looked at with an eye toward providing a complete story of this shutdown. |

| Do I need this? | Item |
|---|---|
| | 1. Create and complete a punch list (if your project warrants the extra step). Note whether there is work that cannot be put off or done under normal maintenance conditions. |
| | 2. End-user acceptance (break out the champagne)<br><br> |
| | 3. Contractor delivery of "As built" or "As Installed" drawings (on CD in standard formats if possible) and other contracted documentation deliverables (such as O & M manuals for new equipment, recommended PM routines, BOM and recommended spare parts). |

| | |
|---|---|
| | 4. If new drawings are not part of the contractor's deliverables, do the drawings in-house. |
| | 5. Get copies of all the bills so that the final costs can be booked and you can determine if the money spent was within budget. To facilitate this work, write a rule into all contracts that all bills have to be tendered within 4 weeks of the end of the shutdown unless prior approval is obtained. |
| | 6. The dispute settlement process was active during the shutdown and now the tough issues must be resolved. Any outstanding extras, deductions, change-orders, problems, mistakes, and damage caused by contractors, need to be resolved in an efficient manner. |
| | 7. Prepare a list of shutdown jobs that have not been started. If these jobs can wait, begin preparation of the next shutdown list or turn them over to the maintenance dispatch function for scheduling. |
| | 8. Check the start up list to make sure all items have been dealt with. Be sure all blinds and locks have been removed. |
| | 9. Insure production is stable and of acceptable quality. If it isn't, study and document all actions so that the next shutdown start-up can go smoother. The goal would be to learn from the mistakes and deficiencies in start-up. |
| | 10. Record status of KPIs that can be measured to this point. |
| | 11. Conduct exit interviews with key contractor employees. Get recommendations for improvements for the future. |
| | 12. The Shutdown Narrative Report should be the last activity directly related to every project. The report should be started early in the project because it is vital that documents be captured when they are generated, even if they were subsequently changed. The report should be completed as soon after completion as possible. |

## Punch list

When you put up a building, the process of commissioning or building turnover from the contractor to the owner is as complex as any ritual. The architect creates a punch list where, if everything is complete and punched, the building will be considered to be complete and final payments will be made. After the work, items on the punch list are completed, and the building completion is declared by all parties. The contractor gets paid according to the contract (a small amount might be held for some period). If there is no agreement then a court declares completion!

A punch list process is very useful for completing the last segments of a project (though it is sometimes the toughest to complete). In maintenance shutdowns the end point may be hard to discern because the project is in reality a series of unrelated work orders. Everyone is running around trying to cram in as much work as possible. Usually the shutdown is over at a particular time, whatever is done or not done. Good language in the activity list will forestall many of the problems.

## Completion of Project paperwork and close out and review meetings.

## Phase 5

Obtain final manuals, as-built drawings, updated planning packages, operator directions or manuals. Decide what to keep and what to discard. Organize lessons learned

## Shutdown Review Report

It is essential to capture what happened so that the same mistakes are not made again! Also collect lessons learned by having different players on the shutdown fill out surveys. . They might be asked: Were job instructions clear? Was the estimate within 15% (look at total hours as well as hours by craft)? Did unusual problems occur on this job? Were tools, parts, or supplies needed that were missed by the plan? Were there delays due to problems with permits, or permissions? Were there delays relating to equipment access? Were there delays caused by lack of craft coordination? Were there any other problems with the plan?

| Do I need this? | Item |
|---|---|
| | 1. Collect O & M manuals, as-built drawings, and other papers, and file them in the appropriate areas and libraries. |
| | 2. Update CMMS database with equipment records for all new equipment. Enter baseline PM routines for new equipment (as a starting point). Update process charts. |
| | 3. Post completion review. This is a debriefing, held right after the project is completed to establish lessons learned in order to avoid their repetition. This review would close out the Shutdown Narrative. |
| | 4. Over the few months after the shutdown, planners should schedule several periods to review the job plans from the shutdown. Details of all jobs should be filed for future reference. Details of any job that is likely to be repeated (either as a normal maintenance job or as a shutdown job) should be cleaned up before filing. |
| | 5. Shutdown Post Mortem Review (three to six months after completion) Costs incurred should be compared with projections. Failure to deliver expectations damages credibility, which appropriately influences approval of future capital requests |

*Thank you for reading.*
*I hope this book helps your next shutdown*

Joel Levitt

# Bibliography, Webiography and Resources


New Standard Institute (NSI) is a seminar provider (both public and in-house) and a publisher of Computer Based Training (CBT). Elements and examples of their information are well distributed throughout this work. I teach some of their Shutdown courses and their Planning and Scheduling courses and learned their approach. They offer CBTs in Maintenance Shutdowns, Turnarounds and Outages and Maintenance Planning and Scheduling. You can contact them at http://www.newstandardinstitute.com/

*Maintenance Planning, Scheduling and Coordination* is a book by Don Nyman and Joel Levitt. It is published by Industrial Press in New York. Hardcover: 320 pages; 1st edition (May 15, 2002) ISBN: 0831131438 Available from Amazon.com or www.Industrialpress.com

One great resource comes out of Canada. PEM is a major maintenance magazine in Canada and publishes handbooks of articles on a particular theme. On the theme of Shutdowns they offer: The Plant Professional's Handbook on Plant Shutdowns PEM (Plant Engineering and Maintenance) a Clifford / Elliot Publication Volume 24, Issue 6 December 2001

Managing Projects by John Reddish, C.M.C. Advent Management Associates in West Chester, PA. This course and workbook has some interesting insights into the field of Project Management.

Project Management Tools and Techniques Training Seminar by Project Management Mentors, San Francisco, CA 1993

*Turnaround Management* by Tom Lenahan is published by Butterworth Heinemann (September 1999). Hardcover: 183 pages, ISBN: 0750642831, Elements of Turnaround management are sprinkled throughout the text. It is an excellent reference book for turnarounds.

Dr Nigel Ellis is one of the world's leading experts on falls and fall protection. He supplied a variety of insights into fall safety that are referred to throughout this book. He can be reached at Ellis Fall Safety Solutions, www.fallsafety.com

A couple of interesting books including *The Story of the Hoover Dam*, which was compiled from stories written in the Ingersoll-Rand magazine, Compressed Air. The compilation was published just after substantial completion in 1935. Another good story was the *Manhattan Project* by Stephane Groueff published by Bantam Press NY. Both books contributed to the history section of Project Management.

www.kidasa.com is a designer of scheduling software and MS Project plug-ins. Their products range from a basic scheduler for $49 to a complete system for under $300. We used their software for the Gantt chart of the Reagan Airport construction schedule.

*Learning from Accidents* by Trevor Kletz. I used this book in the accident investigation section. The author has several excellent books on safety. Other books include *What Went Wrong? Lessons from Disaster: How Organizations Have No Memory and Accidents Recur.* All these books are based on actual accidents (which are clearly dissected for the reader). His books are published by Butterworth-Heinemann www.bh.com and like all the books referenced can be found at www.Amazon.com.

Scheduling Guide for Program Managers published by Defense Systems Management College, Fort Belvoir, VA. 1990

On of the most fun research tools on the Internet is the Wikipedia project, which is a publicly written encyclopedia with articles on a variety of topics. http://www.wikipedia.org/wiki/USS_George_Washington_(SSBN-598))

To see an Internet based free Gantt tool go to http://associate.com/gantt/

http://www.aspenational.com/ is the Association of Professional Estimators. Includes papers on estimating shutdowns and other resources.

The University of Alabama in Tuscaloosa has a Certification in Project Management. It consists of (7) 2-day workshops and a preparation course for the Project Management Professional certificate. http://bama.ua.edu/

http://www.fek.umu.se/irnop/projweb.html WWW-Guide to Project Management Research Sites

I would like to thank the Project Management Institute. We used several resources including PMBOK Guide and Digital Norseman, (http://www.digitalnorseman.com/) Preben Ormen (who works for KPMG LLP regular time) for his useful Project Management advice. PMI was founded in 1969; Project Management Institute (PMI®) has grown to nearly 70,000 members worldwide. PMI establishes Project Management standards, and provides seminars, educational programs, and professional certification that more and more organizations desire for their project leaders. They can be found at www.pmi.org.

http://www.interplansystems.com is one of the leading organizations that help people with turnarounds. They sell support software, specialized project management software, and consulting services. Their eTaskMaker software is extremely useful for setting up activities for an industrial shutdown. They have two free training aids:

Turnaround Project Planning Primer
http://www.interplansystems.com/turnaround-project-planning-primer/toc.html

Turnaround Project Management Primer

http://www.interplansystems.com/turnaround-project-management-primer/toc.html

http://www.msproject.com/ A site dedicated to the Microsoft Project community.

MACRO is a project to help organizations manage projects. The British government sponsored it along with some of the leading contractors and consultants. One of the leaders was John Woodhouse who can be reached at www.twp.co.uk

http://www.guysoftware.com/planbee.htm Is a vendor of a Project management System (cost is under US$100)

http://www.planningplanet.com/ A free source for estimates, information, chat, forums about shutdowns.

http://www.projectmanager.com/ This is a job bulletin board for Project Management professionals.

http://www.projectkickstart.com/html/tips7.htm Some basic Project Management tips from Jeff Crow's web site are useful to maintenance project managers.

http://www.primavera.com/ Primavera is one of the oldest large-scale Project Management software companies. They feature extensive training in maintenance turnaround applications (see the page: http://www.primavera.com/products/pe_maintenance.html).

http://www.projectmagazine.com/ he Is a free on-line magazine dedicated to the practical side of project management.

http://www.pmforum.org This is an association with a web-based library, discussion groups. Good dictionary of PM terms and usages.

http://www.smartdraw.com/specials/projectchart.asp?id=15063 provides Project management software for US$69-$129.

http://www.stylusinc.net/ms_project_tutorial/project_management.shtml
For a tutorial in Microsoft Project

# Glossary

**Activity:** The basic unit of project management with duration, sequence, and resources.

**Actual Finish Date (or time) (AF):** Actual time activity was completed

**Actual Start Date (or time) (AS):** Actual time an activity was started

**Administrative Closure:** The phase of a project where all bills are in and paperwork is completed

**Arrow:** Geometrical representation of an activity in the ADM method, and of a relationship in PDM method.

**Arrow diagramming method (ADM):** Method of diagramming a project where arrows represent activities with the tail being the start and the head being the end. The length of the arrow is not related to the duration of the project.

**As Built:** Drawings that reflect how the asset was actually built, with accurate actual dimensions.

**Backward Pass:** Stepping through a network starting with the completion date and working backward through the different paths. The backward pass results in the (EFT, LFT) latest completion times and is calculated right to left in standard networks.

**Bar Chart:** Graphic method where each bar represents an activity and the length of the bar is related to the duration of the activity (also see Gantt chart).

**Baseline:** Beginning version of project used for comparison (the baseline is saved for future comparison).

**Bill of material (Also BOM):** A list of all parts and materials necessary for a repair.

**Budgeted Cost of Work Performed:** Total budgeted cost including overhead for a unit of work such as an activity.

**Calendar Unit:** Unit of time of chart (hours are used in short shutdowns, days, or shifts in longer ones).

**Change in Scope:** Any change in scope of the shutdown from the baseline project diagram.

**Clearance:** See permit

**CMMS:** Computerized Maintenance Management System, system used to manage maintenance work.

**Chart (or sometimes in shutdowns called code) of Accounts:** Numerical breakdown of project structure, (in accounting it means the chart of accounts codes for charging aspects of the shutdown).

**Communications Planning:** Plan to communication information to all stakeholders and to the project management team. Includes what information, how to communicate, and how often.

**Contingencies:** Part of risk management. Allowances built in to budget or time schedule for unknowns and for mistakes.

**Contingency Planning (also see risk management):** Planning for problems that come up in the project. Alternative methods, materials, vendors, are thought through before the project starts and can be put into action if the problem occurs. (Called risk management in this book)

**Contingency Reserve:** An amount of money put aside for problems that might occur.

**Contract:** A legal agreement that binds both parties. In our interpretation, usually for one party to perform work, provide services, or provide parts and the other to pay for them. .

**Contract closeout:** Physical completion of the work and legal completion of the paperwork for a contract.

**Cost control:** Related to changes in scope. Cost control is control of the costs of a project including the changes to the initial baseline project and mistakes in estimates or bills of materials.

**Cost estimating:** The art and science of calculating the elements of costs and deciding how much is unknown and picking a cost or hours to complete the chosen work.

**Cost Plus Contract:** Contract where the buyer pays all costs plus an agreed upon amount or percentage.

**Cost of quality:** Costs to manage the quality aspects of the job including inspection, training, certification, controls, and rework where necessary.

**Crashing:** Completing the project in as short time as possible.

**Critical Activity:** Any activity on the critical path. Also any activity that has no float or where the earliest finish time and the latest finish time is the same. This usage is different from the common definition of critical. Be careful when using this term outside project management contexts.

**Critical Path:** The shortest timed path through a project where the project is completed. The critical path will change as activities are completed before or behind schedule. The critical path is made up of all critical activities.

**Critical Path Method (CPM):** The project management method developed in 1956 that identifies the critical path and manages the whole project by managing the critical path (and near critical path).

**Deliverable:** A defined output that can be completed, measured, and determined to be finished. Can be a unit of work, or a report.

**Dependency:** For activities, a dependent activity is one that cannot be started until another activity is completed.

**Dummy Activity:** An activity with zero duration and resources just used to resolve logical problems in the diagram.

**Duration:** The length of time it takes to complete an activity. Duration is measured in hours, days, or weeks. Duration may depend on crew size.

**Duration Compression:** Techniques like increasing crew size to reduce duration. Related to fast tracking and crashing. There is a practical limit to compression based on the nature of the work and if it can be divided among more workers.

**Early Finish time (EFT):** The earliest time that the unfinished portions of a project can be completed, based on the logic of the project.

**Early Start time (EST):** The earliest time that the unfinished portion of a project can be started, based on the logic of the project.

**Effort:** Usually an expression of the number of labor hours. Effort divided by the number of workers is duration.

**Estimate:** Analyzing work content and deciding the effort level and duration required to complete it.

**Emergent work:** Work that emerges during a shutdown after openning equipment up. The need for the work might be undectable before disassembly. Exxon calles this work discoverables.

**Execution Phase:** The phase of the shutdown where the repairs begin. The execution phase begins after planning is complete and when it is complete the project moves into the completion phase.

**Fast Tracking:** Performing tasks that are normally done sequentially, concurrently. A fast track project might be being designed in one area while also being built in another.

**Fixed Price Contract:** A contract where the scope of work and the price are fixed.

**Float:** The amount of time an activity can be delayed before it becomes critical and delays the outcome of the project.

**Free Float:** The amount of time an activity can be delayed before it impacts the next activity start time.

**Gantt Chart:** A horizontal bar chart where each bar is an activity and the length of the bar is proportional to the duration. The location of the bar horizontally shows when the activity starts and when it is supposed to be finished. Shading or coloring the bars can denote the present status of the activity.

**Hanger:** An unintentional break in a network usually caused by a missing activity or a missing logical relationship

**Initiation Phase:** The first phase of a shutdown where the lists are drawn up and authorization is given

**Lag:** An intentional delay in a subsequent activity. When you are pouring concrete you might delay the next task a few weeks to allow the concrete to cure.

**Late Finish Date (or time LFD):** (in CPM) The last times you can finish an activity without impacting the milestone or projected finish time.

**Last Start Date (or time LSD):** (in CPM) Last moment that an activity can be started before it will interfere with a milestone or the project finish date.

**Lead:** Advancing a successor task to begin before the completion of its processor task.

**Logic Diagram:** The graphical representation of the activities of a project in proper relationship as to dependencies and independencies.

**Logistics:** The whole field of managing parts, supplies, materials including transportation, storage, and delivery to jobsite.

**Lump Sum:** A contract with a single amount of money paid such as a lump sum of $100,000 to paint this tank. Same as fixed price contract.

**Maintenance days:** Are the number of days the shutdown maintenance and project work goes on. Days for cool down, decontamination, and startup are subtracted to yield maintenance days.

**Master Schedule:** A summary document displaying the major networks of a project. Also might be known as a milestone schedule.

**Milestone:** A significant event in a project usually related to completion of a major deliverable.

**Milestone schedule:** see master schedule.

**Mitigation:** Rendering a risk less potent in its effect on the outcome of a project.

**Mobilizaton:** Contractor assembles people, materials, support, tools and equipment for a shutdown.

**Monte Carlo Analysis:** A simulation of the project is run and a distribution is determined for the probability of the duration for critical activities.

**Near critical activity:** Activities that have only a little float so that any problem would put them on the critical path.

**Node:** The point where activities either are expressed (in PERT) or are related to (as junctions in ADM) depending on the charting method used.

**Path:** A set of connected nodes or activities

**Path convergence:** When two paths come together in an activity.

**Path Float:** see float

**PdM (Predictive Maintenance):** Activities to detect impending failures using sophisticated technology.

**PMS(Project Management System):** Computer software to aid in preparing charts, organizing. Activities, and analyzing resources for project management.

**Percent complete:** An estimate of the amount of work completed in an activity or path.

**Performance reporting:** Collecting information and reporting how a project is doing in relation to the baseline.

**Permit:** Document from operations or safety providing permission to do a specific act such as entering a confined space, opening a valve, or using a cutting torch. Also called clearance.

**PERT:** Project Evaluation and Review Technique- developed in 1956 to manage large projects where there is a high degree of uncertainty.

**Planned Job:** A maintenance job where a planner has evaluated the work, broken it into steps, listed the materials and tools, and thought through all the resources needed to complete the work safely and efficiently.

**PM (Preventive Maintenance):** System to manage activities to extend the life of assets or detect when they will fail. PM inspectors will detect potential jobs that can be considered for the shutdown list.

**PM BOK:** Project Management Body of Knowledge, basic knowledge for successful project management. Identified by the PMI- (Project Management Institute (see resource section).

**Precedence Diagramming Method:** Boxes represent Activities and arrows diagram the logical relationships between the activities.

**Precedence Relationship:** The relationship between two or more activities where one group cannot start until the other group is completed.

**Predecessor Activity:** The activity that must be completed before the next (successor) activity can begin.

**Preparation Phase:** The longest phase of a shutdown where all jobs are listed, analyzed and scheduled for shortest duration and lowest cost, consistent with safe practices.

**Prefabricate and Prefabs:** All work that can be before the shutdown to save time during the shutdown. Could include assembling piping spools, building welded sections, pre-assembly of cabinets, etc.

**Program:** A series of related projects. Commonly used in the military where weapons programs are made up of a series of projects.

**Program Evaluation and Review Technique:** see PERT

**Project Charter:** Permission given by top management for the whole project effort.

**Project Cost management:** An aspect of project management concerned with managing costs and tracking budgets.

**Project communications management:** see communications planning

**Project Human Resources Management:** Systematic methods to make the best use of the people involved in the project.

**Project Life cycle:** The five natural phases of a project (Initiation Preparation, Execution, Termination, Completion).

**Project Management (another use of the initials PM):** The whole field of managing projects.

**Resource:** Anything needed to complete a job including labor, parts, tools, equipment, etc.

**Resource leveling:** Shifting around the time when a resource is used (within its float) to minimize how many resources are needed.

**Risk management:** Processes to reduce, mitigate, eliminate, and manage, all the things that could derail, delay, or disable a project.

**Risk identification:** Determining what events could negatively impact the project.

**Risk response:** Activities designed to mitigate or eliminate events that could disrupt the project.

**S-curve:** Graphic display of costs, hours against time. The curve starts slow, then speeds up, and slows toward the end of the project (so that it looks like an 'S').

**Schedule:** The dates for activities and completion of milestones. Developed after the input of all the activities (work orders) and their durations, resources, and relationships.

**Schedule variance:** Any difference between the actual activities scheduled and baseline scheduled activities.

**Schedule finish date:** Projected finish date based on the logic and assumptions of the master schedule.

**Scope:** The scope of a shutdown is all the jobs, projects, and other work to be done.

**Scope Management:** The scope of a project is the jobs to be done. Managing the scope involves making sure items are not added or deleted from the scope without due process.

**Shutdown days:** are the number of days from the last full good production to the first full good production after the shutdown.

**Slack:** PERT term for float

**Stakeholder**: Any party that has a stake or an interest in the outcome of the shutdown.

**Target completion date:** Completion date that is set as a constraint.

**Termination Phase:** Phase 4 where the plant is turned back to operations.

**Time-scaled network diagram:** Network diagram where the arrow lengths are proportional to duration.

**Turnaround Package:** The complete group of documents describing all work and showing who will do what, when. The package has a schedule, multiple job packages, and all the documents to make the shutdown run smoothly.

**Unit Priced Contracts:** Contracts where the cost is per unit. Unit priced contracts can be as varied as tree planting (cost per tree), boiler tube cleaning (cost per tube), steam trap replacement (cost per trap), etc. Used where the exact number of items is either not known or is subject to change.

**Workaround:** What is done to get around a problem or risk. Different from a contingency in that a workaround is not planned but executed on the fly at the spur of the moment.

**Work Order:** Maintenance document describing a job to be done and providing an authorization to proceed. Also used to collect details of what happened, and with parts used for history.

**Work package:** All the documents for a maintenance job or an activity, forming a detailed description with enough detail to be executable. At a minimum, includes the work order and bill of material.

# INDEX

# Courses Available from Joel Levitt

| Title | Short Description |
|---|---|
| **Maintenance Management Includes *Handbook of Maintenance Management*** | Introduction to maintenance management concepts including benchmarking, TPM, economic analysis, PM, RCM, planning. Can be designed with additional topics |
| **Maintenance Management for Buildings and Facilities** | Complete course to manage maintenance building complexes (airports), venues (stadiums), office buildings, etc. Does not duplicate Maintenance Management |
| **Maintenance Management for Factories and Process Industry Includes *Managing Factory Maintenance*** | Complete course to manage factories, refineries, batch plants. Does not duplicate Maintenance Management |
| **Breakthroughs in Fleet Maintenance** | Complete course to manage truck, car, train maintenance. Does not duplicate Maintenance Management |
| **Management Skills for Maintenance Supervisors and Team Leaders** | Topics include: supervision style, delegation, planning, PM, PPM, basic time management. |
| **Advanced Management Skills for Maintenance** | Topics include: Advanced time management, communications, craft training. Can be added to management skills for a 4-day course |
| **World Class Maintenance available on audio tape** | Talk on the attributes of world-class maintenance. Used by organizations to inspire the maintenance people. Is included in several courses and can be added to others |
| **Preventive and Predictive Maintenance Includes *Complete Guide to Predictive and Preventive Maintenance*** | Complete discussion on all aspects of PM. Longer courses add depth and examples suited to your organization. |
| **Time Management for Maintenance Professionals** | Training on using the little time available to good effect. Includes 4 projects to improve time management and new thought models to improve achievement. Part of supervisor courses and factory courses. |
| **Maintenance Planning Includes *Maintenance Planning, Scheduling and Coordination*** | Complete course and hands-on workshop on planning. Includes all steps in planning work for a maintenance department |
| **Managing Maintenance Shut down and Outages Includes *Managing Maintenance Shutdowns and Outages*** | Complete course and hands-on workshop on planning and managing large maintenance projects and shutdowns. Can be added to maintenance planning to make a full week course. |
| **Internet for Maintenance Management Includes *Internet Guide for Maintenance Managers*** | Course and workshop to set-up the use of the Internet for a maintenance department. All 15 uses of the Internet are demonstrated. Company bookmark project is started. |
| **Optimizing the Maintenance Inventory** | Review of all the issues surrounding running an effective inventory including theory, economics, EOQ, capital spares, rebuildables |
| **TPM (Total Productive Maintenance)** | Introduction to TPM in a factory environment. Can be added to PM course for a complete view of equipment care. |
| **Cost Reduction Project** | A workshop for maintenance craftspeople that introduces cost reduction and how to carry it out. Ideally, the three days are spread over several weeks. . |